ROWAN UNIVERSITY
CAMPBELL LIBRARY
201 MULLICA HILL RD.
GLASSBORO, NJ 08028-1701

SCIENCE IN CULTURE

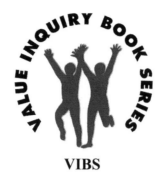

VIBS

Volume 185

Robert Ginsberg
Founding Editor

Peter A. Redpath
Executive Editor

Associate Editors

G. John M. Abbarno
Mary-Rose Barral
Gerhold K. Becker
Raymond Angelo Belliotti
Kenneth A. Bryson
C. Stephen Byrum
Harvey Cormier
Robert A. Delfino
Rem B. Edwards
Andrew Fitz-Gibbon
Francesc Forn i Argimon
William Gay
Dane R. Gordon
J. Everet Green
Heta Aleksandra Gylling
Matti Häyry

Steven V. Hicks
Richard T. Hull
Mark Letteri
Vincent L. Luizzi
Adrianne McEvoy
Alan Milchman
Alan Rosenberg
Arleen L. F. Salles
John R. Shook
Eddy Souffrant
Tuija Takala
Anne Waters
John R. Welch
Thomas Woods
Emil Visnovský

a volume in
Gilson Studies
GS
Peter A. Redpath, Editor

Piotr Jaroszyński

SCIENCE IN CULTURE

Piotr Jaroszyński

Translated from the Polish by
Hugh McDonald

Amsterdam - New York, NY 2007

Cover illustration: Étienne Gilson, painted by Jacqueline Gilson, photographed by Louis-Gildas and Benjamin Morand with technical assistance of Ralph Norden-hold. Used with the kind permission of Cécile Gilson.

175.5
.J37
2007

Cover Design: Studio Pollmann

The paper on which this book is printed meets the requirements of "ISO 9706:1994, Information and documentation - Paper for documents - Requirements for permanence".

ISBN-10: 90-420-2136-5   3 3001 00943 9414
ISBN-13: 978-90-420-2136-5
©Editions Rodopi B.V., Amsterdam - New York, NY 2007
Printed in the Netherlands

To my wife Hanna and daughter Agata

CONTENTS

Part Three

PHILOSOPHY AND THEOLOGY IN RELATION TO REVELATION

Part Four

SCIENCE: TOWARD TECHNOLOGY AND IDEOLOGY

Contents   xi

Part Five

SCIENCE AS TECHNOLOGY

Part Six

THE TRANSFORMATION OF PHILOSOPHY INTO IDEOLOGY

Part Seven

TOWARD A NEW WORLD ORDER

Part Eight

THE PLACE OF SCIENCE IN CULTURE

LIST OF ILLUSTRATIONS

FOREWORD

Science stands beside morality, art, and religion as one of culture's first and essential realms. Many works on the history, methodology, and logic of science exist. But no serious work exists on science as culture has affected it, on how science compares with our understanding of ourselves as personal, as cognitionally responding with an understanding of the things we know.

Piotr Jaroszyński has filled this gap in our understanding of science as a basic realm of human culture. He has presented the Greek context of scientific knowledge from the period of classical culture. The kind of philosophical thought that arose there has developed over time and remains to this day. His philosophical thought is rational and verifiable. The conceptions of scientific knowledge developed within Ancient Greek philosophical thought. In the face of facts, the leading question was to ask "why?" This type of knowledge was ordered to theoretical knowledge (*theoria*) as the contemplation of truth. The aim of science was truth, for the truth alone could guarantee our normal development in culture's other realms: morality, art in the broad sense, and religion.

Over time science's unique purpose yielded to another: to serve utility and increasing technical management of life. Heretical gnostic thinkers contributed to the departure from the search for truth. Their influence increasingly obscured scientific knowledge as a contemplation of the truth. New philosophies that presented human beings in merely subjective terms furthered this process. In subjective philosophy human beings became "creators of the truth," not personal beings who "seek the truth." Subjectivization of scientific knowledge led to a forgetting of the scientific question of "why?" The new question was "how?" As science looked at the world in these new terms, the idea of who we are became distorted.

Jaroszyński's presentation is unusually clear and erudite. He has studied the original sources and formulations of philosophers and other intellectuals who created revolutions in thought. Among other things, he presents discoveries about magic's history, the Cabala of pseudo-mystics and ideologues, anti-metaphysical and anti-religious attitudes in the construction of the conception of science as part of a program to establish a "new world order"—*novus ordo rerum*—by an elite of initiates who would be joined together by gnosticism, pseudo-mysticism, and ideology that would seek great social reforms.

We need this work to make us aware of the present state of science. It will benefit those who study this topic.

Mieczysław Albert Krąpiec
The Catholic University of Lublin

PREFACE

When we create science as knowledge, we carry out the most essential requirement of our nature. Science as a knowledge of the truth is a fully autonomous value common to all human beings. It does not need to be justified in terms of other, such as practical, ends. Science as truth is an autonomous value and the basis for the other essential values.

The other essential values that also aim at satisfying our essential desires and laws of development are the desire for the good and the beautiful. When the desire for knowledge guides us, we produce science: we discover the truth. When the desire for the good guides us, we create ethics: we discover moral principles. When the desire for beauty guides us, we produce art.

These three directions in human action are closely bound with our human essence because they are all necessary conditions in our nature's realization. They result in the three major components of culture: (1) science as the bearer of truth, (2) ethics or morality as the bearer of good, (3) and art as the bearer of beauty. All three together form the rational component of our nature. They are a product and expression of human nature. But science has first place because knowledge is the first aim of human reason, and truth is the essential criterion for evaluating ethics and art.

This book presents this philosophy of science, its genesis, essence, and history. It is a defense of this philosophy of science in its confrontation with other conceptions, such as pragmatism, anarchism, and different kinds of subjectivism that have departed from the autonomy of science as knowledge (as the bearer of the truth) and have separated it from the nature of human reason. These alternative conceptions have led to the degradation of (1) science as a component of culture, (2) culture as such, and (3) human beings.

Piotr Jaroszyński's book is the first work dedicated to the position of science in culture. It fills an crucial gap in the literature on science and culture. It is (1) a valuable, outstanding, book because of the importance of the problems it addresses and Jaroszynski's profound grasp of these problems; (2) a clear presentation with cogent logic and beautiful language; (3) historical and systematic.

The author clearly defines his position, which serves as a foundation for an insightful critical evaluation of other positions in the philosophy of science and different ideologies. He shows how they have brought destruction to human beings and culture throughout history. I would like to emphasize that this book contains crucial new discoveries.

As a beautiful scholarly work, this book warrants publication on its own merits and for its didactic value. I have no doubt that it will be widely read and will bring honor to the publisher.

Tadeusz Kwiatkowski
Lublin, 2006

INTRODUCTION

Many of us think that science is a monolith that grows, becomes more perfect over time, and leaves behind what is obsolete. We might think about how, in physics and chemistry, the theory of matter has changed so profoundly. We might recall the opinions of Ancient thinkers that appear to be a relic of the past. Today, science develops quite quickly, becomes increasingly specialized. Its changes become increasingly difficult to follow.

The average mortal sees scientific progress in terms of use of technologies that would not exist without science, not by a study of theories. Technology's new products provide constant witness to science's constant progress.

Not so long ago we often heard that scientific and non-scientific world-views exist. Marxism claimed the scientific world-view as its own. It associated the non-scientific world-view with Christianity. So Christianity had to be rejected. As we take a closer look at the different controversies surrounding the conception of science, we will see that many such controversies have existed, and still do. Science is no universally recognized monolith. And it has generated bitter disputes that go beyond science to questions of world-view, ideology, politics, economics, religion, and even civilization in general.

While science has brought enormous technological benefits to us, science may be the object of extra-scientific controversies about the role that a specific conception of science plays for some world-view, ideology, religion, or civilization. The controversy about the conception of science is no mere methodological or historical dispute. It touches on civilization.

For someone not acquainted with these matters all the arguments might appear to concern methodology or history. Someone rejects some conception of science in the name of some criterion of scientific knowledge, while the same criterion has been formulated on the basis of some ideological premises. An anti-religious ideology will strike at any concept of science that recognizes religion as something rational and meaningful. Controversies between religions or between sects in the same religion may also involve different conceptions of science.

Science is too influential and powerful a domain in culture and civilization to be left to its own devices. Science and the conception of science may become tools for different reasons and different aims. This side of the coin is not well known today and should be brought to light.

Someone may object that this presentation will involve a context foreign to science and that it will be part of a controversy outside of science. This objection is not so. We may identify the founders of science as such when science possessed its own immanent goals. We may also show the moments when the conception of science changed under the influence of factors external to science.

While everyone knows that discoveries in civilizations predate the Greeks, we may speak with historical justification about the start of science as an established domain of culture only in Greek civilization. When people deny that the

Greeks were the first people to discover science, as happened even in Ancient times, they are not concerned with historical accuracy. They are reading history ideologically in terms of controversies between nations, religions, and civilizations (for example, between the Greeks and Egyptians, or the Greeks and the Jews). Later controversies about the conception of science also contain the same non-scientific element, and this exacerbates the controversy, for example, in the Protestant critique of Aristotelianism. In one way or another, we should consider that the controversy over science or the conception of science is deeply rooted in controversies extrinsic to science.

In this work, I would like to (1) bring to light the non-scientific contexts of the controversy over science and (2) show the influence of science in culture (how the extra-scientific context influences the conception of science, and how science influences culture and its particular domains). Doing this will be interesting in view of the wealth of historical material and because the problem is still relevant. Science has been brought into the orbit of ideological conflicts. Politicians, business people, and journalists call upon the name of science. While science contributes to the advances of technology, it also poses threats to human beings and to the Earth. Scientific methodology and science as such have no necessary competence to show the place and role of science in culture. This is the job of the philosophy of culture.

ACKNOWLEDGMENTS

I am grateful to the many people and organizations that played a role in this project in so many different ways. I cannot acknowledge them all here. But I especially want to recognize the following: Mieczysław Krąpiec, Peter A. Redpath, Hugh McDonald, Roman Czyrka, Jadwiga Kądziela and the *Servizio Fotografico de "L'Osservatore Romano"* (for permission to publish photographs taken at the Vatican), and *Polonia* from Calgary (Canada).

Part One

THE RISE OF THEORETICAL KNOWLEDGE

Mieczysław Albert Krąpiec, Lublin School of Philosophy

One

THE DISCOVERY OF SCIENCE: GREECE OR THE EAST?

To this day, no one has properly investigated the question of whether science was born in the circle of Greek culture or was already known earlier to the Egyptians or Babylonians. The Ancient Greeks formulated this question. They thought that they were the first. But, even during the Hellenistic period, and later, during the Middle Ages, the dominant view was that wisdom came from the East, while the Greeks were simply talented students.

Today some people try to disparage the Greek's contribution to Western culture. Their critiques appear more plausible because, with the contemporary decline in classical education, few people have adequate knowledge of Greek and Latin. The average person has quite a feeble idea of the influence of Greece upon Western civilization. Sophisticated people show a special interest—often mixed with snobbery—in the East as the source of wisdom and power. Some apparently think that despising the Ancient world displays good taste, and that we should avoid it as the source of distortions in the Christian religion.

To establish what nation was the cradle of science is crucial. It weighs upon the shape of our civilization. Who can seriously deny that the broad development of science is specific to what we call the West? If science came from the East, we must change the way we think about the West. Was Greece or the East the author of science as a specific cultural domain?

The Greeks did not try to hide their debts to other peoples. Herodotus wrote that the Greeks received the cult of Dionysus, the names of almost all their gods, and their belief in the transmigration of souls, from the Egyptians.[1] Plato was full of admiration for the Egyptians. In the *Timaeus*, he relates how Solon wanted to display his knowledge to the Egyptian priests and started to speak of the earliest history he knew. A priest interrupted him and said that the Greeks were still quite young, and their memory did not go back very far. The Greeks had no old opinions passed on by Ancient tradition or any science that was faded with time.[2] The Greeks had borrowed many of their views about the past from the Egyptians.

Aristotle wrote that the mathematical arts first developed in Egypt. "Hence when all such inventions were already established, the sciences which do not aim at giving pleasure or at the necessities of life were discovered, and first in the places where men first began to have leisure. This is why the mathematical arts were founded in Egypt; for there the priestly caste was allowed to be at leisure."

Plato also wrote of the Egyptian origins of mathematics: "I heard that at Naucratis in Egypt there was one of the Ancient gods of the country, to whom was consecrated the bird known as the ibis. The god's name was Theuth. He first invented numbers, arithmetic, geometry and astronomy. He also invented draughts and dice, and most importantly, the alphabet."[3]

Eudemos of Rhodes said that Thales, and Isocrates said that Pythagoras, had acquired their knowledge of mathematics from Egypt. Democritus allegedly traveled to India and learned much there.[4]

Aristotle mentioned the Babylonian roots of astronomy. He said: "For we have seen the moon, half-full, pass beneath the planet Mars, which vanished on its shadow side and came forth by the bright and shining part. Similar accounts of other stars are given by the Egyptians and Babylonians, whose observations have been kept for very many years past, and from whom much of our evidence about particular stars is derived."[5]

In many respects the Greeks were in debt to older civilizations, and they loyally acknowledged this. But did the Greeks produce nothing new?

The Greeks had an original concept of science with respect to the particular or specialized sciences and philosophy, the queen of the sciences. Science became a distinct domain of culture for the first time in Greece.

In his classic work, *Early Greek Philosophy*, John Burnet made a close examination of Egyptian mathematics and Babylonian astronomy.[6] He analyzed the Rhind Papyrus in the British Museum and concluded that the mathematical problems it treats are exclusively practical in character. For example, it tells how to weigh corn and fruit, and how to divide a specific number of measures among a determinate number of people: how many jugs of beer are contained in a particular measure, the wages due to workers for a some part of work, and so forth. The Greeks called these skills logistics (*logistike*), which we may translate as a practical skill in correct reckoning.[7]

Plato also thought that every free person should learn this kind of reckoning, as children learned it in Egypt. He says that the Egyptians invented arithmetical games for children to learn with pleasure and amusement. One such game involved distributing garlands and apples. Another utilized the same number to refer sometimes to a larger, then a smaller, number of people. A third consisted of teaching numerical order by arranging boxes or pairing wrestlers together. A fourth involved distributing cups of different kinds (like gold, silver, brass), sometimes intermixed, sometimes not.

By adapting numbers in common use to the personal amusement of the children, Plato maintained that the Egyptians were able (1), in military matters, to "make more intelligible to children the arrangements and movements of armies and expeditions"; (2), "in the management of a household they make people more useful to themselves, and more wide awake"; and (3), "in measurements of things which have length, and breadth, and depth, they free us from that natural ignorance of all these things which is so ludicrous and disgraceful."[8]

If someone lacked this kind of knowledge, Plato claimed that person would no longer resemble a human being: "O my dear Cleinias, I, like yourself, have late in life heard with amazement of our ignorance in these matters; to me we appear to be more like pigs than men, and I am quite ashamed, not only of myself, but of all Hellenes." And Marcus Tullius Cicero approvingly recalls an anecdote about "Plato or some other philosopher":

> When a storm cast him from the deep sea on to the deserted shore of unknown lands, the others were afraid because they did not know where they were, they say that he noticed some geometric forms drawn in the sand. When he saw these, he cried that they should be in good spirits, for he was seeing traces of men. He did not say this because of the cultivated field that he saw, but because of the signs of learning.[9]

Before the Greeks, logistics was not yet true arithmetic (*arithmetike*). What we learn in school under the name of arithmetic is more like the Greek logistics. What the Greeks called *arithmetike* is the scientific investigation of numbers.[10]

Egyptian geometry was also practical. Herodotus wrote that the Egyptian geometry was primarily about the division of fields. Every year the farmers had to pay a tax to the Pharaoh based on their land holdings. The matter was complicated when the Nile would rise and destroy part of a farmers' plot and his rent would have to be lowered accordingly. Herodotus said:

> This king also (they said) divided the country among all the Egyptians by giving each an equal parcel of land, and made this his source of revenue, assessing the payment of a yearly tax. And any man who was robbed by the river of part of his land could come to Sesostris and declare what had happened; then the king would send men to look into it and calculate the part by which the land was diminished, so that thereafter it should pay in proportion to the tax originally imposed."[11]

Herodotus concluded: "It seems to me that in this way they invented the geometry that came thence to Greece."[12]

Building pyramids required higher skills. But these skills were still of a practical character. The builder would have to find the number that expressed the relation between the hypotenuse and the base of a triangle. He did this empirically, dividing half of the diagonal by the length of the base. This is a typical example of the extent of Egyptian geometry.

The Greeks characteristically took a theoretical approach to geometry. They created their own terminology, and our mathematical terminology to this day reflects its Greek origins.[13]

As for Babylonian astronomy, the Babylonians definitely observed the Heavens and kept records of the fixed stars, especially those in the constellations of the Zodiac. They gave proper names to the stars and described their apparent motion. They knew the times of the solstices and equinoxes, and used ellipses to calculate how long a star would take to return to a certain point. All these observations were for one purpose: to foresee the future.[14] Babylonian astronomy was really a form of astrology.

The Babylonians linked their astornomical observations and astrological beliefs to teachings they already held in the time of Hammurabi (the eighteenth-century B.C.) and that we can read on their clay tablets. They believed that the gods pass on information to people on Earth by different phenomena in the Heavens and on the Earth.

The lunar eclipse was one of the first events they treated as a celestial omen. Later they treated the first and last appearance of the Moon, planets, and individual stars as signs.

Ernan McMullin writes that the Babylonians were convinced that their ability to interpret these signs let them foresee what would happen on Earth, events such as political changes, unsuccessful harvests, and so forth.

The key to the interpretation of different phenomena remained unchanged for many centuries, and it was the canon *Enuma Anu Enlil*. The following rule is an example of astrological associations: "If the stars of the True Shepherd (Orion) twinkle, some important person will become powerful and perform evil deeds."[15]

Increasingly accurate astronomical observations started in the sixth-century B.C., while the great era of Babylonian astronomy was in the third-century B.C., already after the fall of Athens.[16] Babylonian astronomy became scientific under the influence of the Greeks.

Just as the Egyptian knowledge of mathematics was merely a form of reckoning, so Babylonian knowledge of the stars was merely astrology. It was not yet a science. It was a very narrow and static body of knowledge subordinated to practical ends and tied with certain *a priori* religious assumptions about sooth-saying.

When the Greeks became interested in acquiring knowledge, mathematics and astronomy rose to quite a high level. Along with this, new sciences appeared and a rapid growth in science in general occurred.

Logistics, reckoning, became arithmetic. Land surveying became geometry. And astrology became astronomy. The Greeks also developed ethics, politics, geography, logic, rhetoric, poetics and philosophy.

Diogenes Laertius (third-century B.C.) reports that some people say that philosophical study first developed among barbarian peoples. Among the people he lists as reported being philosophy's originators are Magi among the Persians, Chaldeans among the Babylonians or Assyrians, Gymnosophists among the Indians, and Druids or Holy Ones among the Celts and Gauls. He strongly disagrees with this claim. Without question he attributes to the Greeks the start of philosophy and the human race:

There are some who say that the study of philosophy had its beginning among the barbarians. They urge that the Persians had their Magi, the Babylonians or Assyrians their Chaldeans, and the Indians their Gymnosophists; and among the Celts and Gauls there are the people called Druids or Holy Ones . . .. But those authors forget that the achievements they attribute to the barbarians belong to the Greeks, and that not only did philosophy begin with the Greeks, but the human race itself.[17]

Two

WHY THE GREEKS?

Finding one definite reason for the appearance of theoretical, or speculative, knowledge (*theoria*) as a new domain of culture is difficult. The other civilizations were of a sacral and practical character. In these civilizations knowledge was completely immersed in mythology. The two were united, not distinguished. Knowledge served the ends of individual and social life and resembled practical skills more than knowledge considered as such. Among the Greeks, theoretical science arose as a sphere distinct from mythology and art. What were the circumstances of the appearance of "free thought," of thought seeking the truth for its own sake?

Theoria started in Ionia, a Greek colony in Asia Minor. From that culture grew the independent culture of the Greeks, whose center later became Athens. Three cultures existed in and around the Egyptian Sea in the second millennium B.C.: Cretan, Minoan, and Helladic.

Mycenean culture, a variety of the Helladic culture, was close to Cycladic culture. Around 1200 B.C., the Dorians from the north reached Pelopponesia, Crete, Rhodes, Cos, and the shores of Asia Minor. In the ninth-century B.C. the Ionians from the north pushed the Dorians south. The Ionians then took control of the coast of Asia Minor.

Greek Culture was born in Asia Minor, not in Athens, in the late ninth- and early eighth-century B.C.. The first great works of this culture were Hesiod's *Works and Days* and Homer's *Iliad* and *Odyssey*.

The Ionians had a feeling of individuality and independence. They loved to travel, especially to Egypt and the lands of the Near East. They were keen observers. Besides Homer and Hesiod, some other Ionian figures who lived during at least three centuries of that civilization were Thales, Epimenides, Pherecydes, Archelaus, Anaximander, and Anaximenes. Only Empedocles of Acragas was non-Ionian.

Aristotle observed that the Greeks were psychologically opposite the other peoples of Asia and Europe:

Those who live in a cold climate and in Europe are full of spirit, but wanting in intelligence and skill; and therefore they retain comparative freedom, but have no political organization, and are incapable of ruling over others. Whereas the natives of Asia are intelligent and inventive, but they are wanting in spirit, and therefore they are always in a state of subjection and slavery. But the Hellenic race, which is situated between them, is likewise intermediate in character, being high-spirited and also intelligent. Hence, it continues free, and is the best-governed of any nation, and, if it could be formed into one state, would be able to rule the world.[1]

Historians of culture say that the Ionians were a society with no central state, organized by tribe around courts. In Homer's time no monarchy yet existed. The rationalization of mythology played a crucial role in the emergence of *theoria*. We can see the rational treatment of mythology in Homer, and especially in Hesiod. The Ancient Greeks interpreted mythology first in theological, then philosophical, terms.

Aristotle describes Homer and Hesiod as the *protoi theologesantes* (first who reasoned about the gods).[2] Instead of being content to repeat the old myths, they were the first theologizing thinkers who tried to approach the question of the gods in a rational way, with the help of rational thought (hence *theos*—god, *logos*—reason).

They still provided no sufficient rational justification for their views. The rational justifications of these first theologians were mythological considerations (*mythikos sophizomenoi*). They cloaked philosophical thoughts in mythology.

Werner Jaeger notes that Hesiod rationalized the myths more than did Homer, since Hesiod searched for coherence among the myths.[3] Hesiod's work sought to reveal the origin of the gods and the generation and order in the physical universe:

His work shall reveal the origin of all gods now reigning upon Olympus; he will also tell us how the world has come to be, with all its present order. He must, therefore, record all the relevant myths and show how they fit together; he may perhaps have to eliminate many versions that strike him as incorrect, or devise new connections where tradition has not supplied them.

In short, Hesiod (1) presented a genealogy of the gods, (2) explained how the world arose, and (3) showed why the world is now in this state.[4] Hesiod provided a theogony (the origin of the gods) and a cosmogony (the origin of the world). His work was a crucial prelude to the rise of philosophy.[5]

While the religions of Egypt and Babylon were centralized, the religion of the Ionians was not. The Ionians had no priestly caste that would lock away its knowledge for itself and make it an instrument for domination. They had no state religion that would be imposed on all from above.

Orphism was the religion that had the greatest influence on their philosophy. And they regarded Orpheus, the mythical singer, as its author. Orphism influenced philosophy with its ideas, especially the immortality and transmigration of the soul, and its attitude that religion could be a search for truth for its own sake.

Orphism influenced the Pythagorean school. And the Pythagoreans were renowned for their investigations in mathematics and philosophy. While the Orphic movement was esoteric and only for initiates, within it scientific investigations had religious approval. Their discoveries were not kept secret for long.

The achievements of the Pythagoreans became widely known, and they influenced many philosophers, including Plato. These historical realities lead us to another question: What was the essence of the Greek miracle?

Three

BIOS THEORETIKOS

I could translate the term in this chapter heading, "*bios theoretikos*," as "the theoretical life." But this translation would be misleading. Today we commonly think of "life" as the vegetative processes universal to all living things, such as nutrition and growth. The word "theoretical" may suggest hypothetical and impractical speculations, divorced from reality. What could "life" have in common with "theoretical thought"?

The Ancient Greek did not think of life simply as vegetative functions. Life also took other higher forms, including sensing and rational life (life's highest form). Vision, hearing, smell, touch, and taste are an animal's most alive sense activities. And beings that possess these senses are more alive than those that lack them.

Life is greater in a bird and in animals in general than in plants. Human beings have vegetative and sensitive life and the life of reason, freedom. Reason's life is higher, more intensely alive, than vegetative and sensitive life. And we still rightly regarded it as life, since what is not alive has no ability to understand.

We derive the term "*theoretikos*" from the verb "*theaomai*," "to look at" or "to view." The *bios theoretikos* does not consist in speculations divorced from real life. It consists in the contemplation, rational viewing, of reality. This is the highest manifestation of life. Hence, Plato states, "[T]he rich, brave, and wise man alike have their crowd of admirers, and as they all receive honor they all have experience of the pleasures of honor; but the delight which is to be found in the knowledge of true being is known only to the philosopher."

Werner Jaeger adds:

Perhaps what is most characteristic among the merely human features of these first philosophers (who were not yet called by this Platonic name) was their specific spiritual attitude, their complete dedication to knowledge, and their immersion in contemplation, which to the later Greeks (but also certainly to their contemporaries), seemed completely unintelligible, yet evoked the highest admiration.[1]

The Greeks found the life of the reason in human beings to be life's highest manifestation. Other civilizations failed to see a human life that pursues knowledge for it own sake to be the highest kind of human life because they subordinated knowledge to practical ends, such as surveying fields or building tombs. Simply to look at things with understanding for the sake of knowledge and understanding did not occur to them. Recognition of a human being as *bios theoretikos* was Greeks' most original discovery. This is why so many branches of science arose among the Greeks.

In his youthful work called, *Protrepticus* [*Exhortation to Philosophy*] Aristotle urged Themistos, King of Cyprus, to take up philosophy. Aristotle presented the King arguments, including one based on "nature's intention." Today we value science and learning for their practical benefits, and we treat science as a profession. But Aristotle approached the meaning of science by an analysis of our human nature. If we are a product of Nature, then we are the foremost, the highest, of all nature's beings. Nature's products characteristically act toward ends specified by their natures: natural ends.

Evidence of this "natural end" appears most often in the last phase of development. First comes the less perfect. In the course of change it becomes perfected, reaches maturity, as it approaches its own natural best state. Aristotle based this conclusion upon the observation of natural processes. The chick breaks out of the egg before it even knows how to walk, fly, or obtain food for itself. It acquires these necessary skills and perfections over time.

Aristotle observed that, when we examine the process of individual human development, we see that the bodily part matures first and the spiritual part matures later. Wisdom (*phronesis*) appears at the end of our lives, if at all, because we do not become wise all at once. In this sense, as a perfection, the last is the best and is development's end or natural aim. So, Aristotle concluded: "[A] certain form of wisdom is our purpose by nature, and the exercise of wisdom is the final activity in view of which we have come to be. It is therefore clear that since we have come into being in order to exercise wisdom and to acquire knowledge, we also exist in this end."[2]

This powerful formulation concerning the human life's natural end is no arbitrary or wishful thinking on Aristotle's part. It is the result of an objective analysis of nature.

In this line of thinking Aristotle was solidly in the tradition of the first philosophers, such as Pythagoras and Anaxagoras, who said that we were made by God to acquire knowledge and look at things (*gnosai kai theoresai*). We should subordinate the other spheres of human life, including morality, to wisdom. "Therefore," he said, "other things are done in view of the goods that exist in man himself. The things that are a good in the body are done for the sake of the good of the soul. We should develop moral perfection for the sake of wisdom, for wisdom is the highest end."[3]

Aristotle still did not solve the question of what should be wisdom's object. As he tells us, his concern was to establish the supreme position that knowledge for the sake of knowledge has in human nature: "[I]f wisdom is the highest natural end, exercise in wisdom would be the best thing of all."[4] Knowledge for the sake of knowledge is human nature's natural end.

We are composed of body and soul. Two different faculties exist in the soul. The highest living faculty is the reason. The body exists for the soul's sake. And the particular faculties in the soul exist for the reason's sake.

Likewise, the things that happen outside of us exist our sake. And our immanent bodily activities exist for the sake of the soul's operations. The soul's operations exist for the sake of its highest operation: reasoning.[5] Aristotle returns to this same line of reasoning in subsequent fragments and appears to delight in its coherence, which is in agreement with nature and with logic.[6] Hence, he says:

All Nature as possessing reason does nothing in vain, but always toward some end; it rejects the fortuitous and is concerned about the end to a greater degree than the arts, for as we know, the arts are an imitation of nature. Since man by nature is composed of soul and body, the soul is better than the body, and that which is worse must always be the servant of that which is better, and so the body must exist for the sake of the soul. If we suppose that the soul has a rational part and an irrational part, and that the irrational part is worse, we may infer that the irrational part exists for the sake of the rational part. . . . All other things are desired by people for the sake of thinking and the reason, since all other things are desired for the sake of the soul, and the reason is the best part of the soul, and therefore all things exist for the sake of that which is the best.

Aristotle presents another interesting argument for the value, and sake, of knowledge. When we speak today of freedom, we most often think from a subjective point of view, of the possibility of choice. When Aristotle speaks of freedom and servitude, he speaks from an objective viewpoint. When something by its nature exists to serve something else, does not exist for its own sake, it is not free. For example, an axe exists for the sake of the woodcutter who uses it to cut wood. So the axe is not free. By nature it serves the woodcutter's purpose. It is like a servant. It stands objectively lower than the person it serves.

In this context, free thought must be more valuable than servile thought. Free thought serves nothing else. So, it is *theoria*, looking at things with the addition of understanding. Utilitarian knowledge stands lower, is not free, and is no end in itself.

In this way Aristotle put *theoria-sophia* (*sophia*—wisdom) on a pedestal as the most highly valued activities that we perform.[7] A trap exists in the too urgent question of benefit because, at some moment, something must be an end in itself, and it will not derive its goodness from being useful for some other end. Usefulness or benefit cannot be the final criterion for the good because that good considered in itself is beyond, and determines, benefit. Aristotle tells benefit is for the sake of just such a good. The good does not exist for it. To expect otherwise is ridiculous.

It is ridiculous to require from each thing a benefit beyond the thing itself and to ask "what is the benefit of this for us" and "what is the profit." For in truth, we assert that he who asks about this is not in the least like the person who knows beauty and goodness, or he who distinguishes the cause from the secondary cause.[8]

The first philosophers did not value human reason in its operation of theoretical knowledge, and none of them would have agreed with the statement of the youthful Aristotle: "If a man is deprived of the ability to perceive and deprived of the reason, he becomes like a plant. If he is deprived only of reason he becomes like an animal. However, a man who is fully endowed with reason becomes like God."[9]

Human reason is something divine in the human being, and so the life of the reason is for the sake of knowledge considered in itself: *Bios theoretikos* is most valuable, most divine, of all. Aristotle based this conclusion upon objective reasons that we recognize when we analyze nature. These are no suppositions or preferential options. They are a cold diagnosis.

No benefit, pleasure, or any other faculty can compare to the greatness of the good of speculative, or pure, knowledge. In the final fragment of the *Protrepticus*, Aristotle writes: "For reason is God within us" (Hermotimus or Anaxagoras said), and "mortal life includes a part of the god himself." We should either take up philosophy or bid farewell to life and leave this world, since everything apart from it is idle talk (*phluaria*) and babbling (*leros*)."[10]

Plato was aware of a difference between the Greeks and other peoples in their respective approaches to knowledge. He relates in the *Republic* that only the Greeks are *philomathes* (lovers of the sciences), whereas the Thracians and Scythians (peoples of the north) were best described as *thymoeides* (inclined to anger and war), while the Phoenicians and Egyptians were *philochrematon* (lovers of money).[11]

Without the *bios theoretikos*, human life loses its most profound meaning. This is the message of Greek culture. We cannot find this message in any other civilization of that time or before. The rise of philosophy and the particular sciences, and the high social status accorded to them, were a response to the Ancient Greek discovery of the rank of theoretical reason in human life.[12]

Four

PHILOSOPHY'S RISE FROM
SENSATIONS TO WISDOM

The term "philosophy" means a love of wisdom. Philosophy's actual practice predates the term. When the pre-Socratics spoke of philosophy, they used terms such as "*historia*" (investigation) or "*sophia*" (wisdom). The first philosophers were investigators and sages. Their position in society was not yet very high because when philosophy started philosophers had no dominant role in society. That role belonged to poets, lawmakers, and statesmen. Only later did the sage acquire general respect.[1]

Aristotle started his *Metaphysics* with the famous words: "*pantes anthropoi tou eidenai oregontai physei*"—or in Latin, "*omnes homines natura scire desiderant*"— "All human beings have an innate desire to know." By using the word "*physei*" (from "*physis*"—"nature"), Aristotle emphasized that the desire for knowledge is the most natural or innate desire, and this fact can be observed in all of us. He was not thinking only of highly specialized aspirations for learning in the sciences, because only some people have such aspirations. He was thinking of the process of knowledge that occurs in every conscious person. We all naturally activate our powers of knowledge: eyes, ears, touch, and understanding.

Some kinds of knowledge are common to human beings and other animals, while some are proper only to human beings. All animals have the ability to gather sense impressions. And this is what marks animals as different from plants. Plants possess only a vegetative life, while animals also have a sensitive life. Some animals possess memory and imagination and are, thereby, able to grasp things and learn. Only rarely do animals acquire experience.

In us, this process of knowledge accelerates, as it were, at the point where it is completed in other animals. Aristotle wrote: "They all [animals] live by mental images and memory, and in a small degree share in experience, but the human race lives by art and reasoning (*techne kai logismois*)."[2] We share with other animals the ability to collect impressions and remember them. But we differ from animals because we are much more able to profit from experience, and experience generates other kinds of knowledge that animals do not possess: art and intellectual knowledge.

When Aristotle mentioned art, he was not necessarily thinking of the fine arts. The Greek expression "*techne*" was wider in scope and applied to all skills in producing things by employing memory and knowledge. The physician, for example, possesses the art of healing. Aristotle said:

Art arises when from many grasps of experience one general grasp is produced which refers to similar instances. This is merely experience, since it states that something helped Callias in the case of such and such an illness, and also helped Socrates or someone else in particular cases. The fact that in a particular ailment a particular treatment helps all who are defined by one concept, for example, phlegmatics or cholerics, when they are suffering from a fever, is art.[3]

Experience is individual. Art has the essential element of generality: art is a knowledge of what is general. Intellectual knowledge (1) is general and (2) provides a knowledge of causes. A person who possesses such knowledge knows causes. The passages between experience, art, and knowledge are fluid. But, in each case, a new element appears. Wisdom is knowledge's crowning point: a knowledge of causes, especially ultimate causes.

The wise person knows more than impressions, does more than gape at things, and possesses more than the ability to feel whether things are warm or cold. Someone who can only produce something on the basis of such rudimentary experience is not wise.

Wisdom requires knowledge of causes, and the wise person must know the answer to the question "why." While experienced people know only that something takes place, wise people understand why it takes place. To be wise is to understand, and know causes.[4] Also, the wise person knows more than just any causes: the causes of everything (*to panta*). The wise person should know the causes of everything, and in this will be different from the specialist.

Let us examine more closely wisdom's components. The Greek question was *dia ti*? We can translate this as *by what, through what*, or *on what account*? Aristotle states that our perceptions grasp what is particular. For example, that fire is hot; but no perception can explain why fire is hot.[5] Likewise, we can go through the entire field of sense impressions—visual, auditory, olfactory, gustatory, and tactile—and, in each case, we have the same situation: we know or feel that something is so; but we do not know why it is so. Since we do not know why, we cannot say that we understand. Instead we say that we see. To understand, we must know why something is. Only when we learn the causes do we become wise.

Three Greek terms denote causes: "*arche*," "*aitia*," and "*stoicheion*." Good reason exists to retain these Ancient terms. Today, we often formulate the question of causality quite abstractly and speculatively, as if the question had been filtered through many varieties of idealism, especially German, which has weighed heavily upon the philosophical mentality of our times. The first question of Greek philosophy, of causality, is most relevant. And we base this relevance upon its rationality.

Finding a term in English that could render the double meaning of this Greek term is difficult. *Arche* means (1) the principle from which something arises, and (2) the start of generation. These two meanings appear to reveal the two connected aspects of *arche*: (1) the principle from which something arises is, simultaneously, (2) the start of generation. So, initially, *arche* is simply the primordial principle.

The first philosophers, who were called "physicists" or the "Ionian philosophers of nature," started their investigations with the question about this primordial principle from which everything (*to panta*) arose. When they would learn the primordial principle, in some way, they thought they would know everything and could be regarded as *sophoi* or wise men. According to Hippolytus and Simplicius, who drew upon the work of Theophrastus, the term *"arche"* appeared for the first time in a philosophical sense in the work of Anaximander, not Thales.[6]

The question *"dia ti?"* differs from particular knowledge and primarily concerns the primordial principle: the *arche*.

The term *"aitia"* is most often translated as "cause." The term arose from the judicial tradition, as a word bearing legal meaning, and concerning accusation, guilt, and responsibility. The Greeks spoke of *aitia* when someone was accused, found guilty, and had to take responsibility for his actions. *Aitia* also concerned the more concrete situation where someone had been accused and someone had to investigate whether the charges were justified. The person investigating needed a statement of the causes that would be the basis for pronouncing the person's guilt.

In this context, the term *aitia* means more a reason than a cause. Once the fact of the crime has been established, we search for the reason for what took place, to assign guilt, condemn, or exonerate the person accused. The reasons for which we start a judicial process must be real, cannot be abstract. And the whole judicial process exists for the purpose of discovering these reasons. When the philosophers took the word *"aitia"* for their own use, they used it for the reason or reasons that explain definite facts about being or the fact of being as a whole, of everything (*to panta*).

We may translate the term *"stoicheion"* as "element" or "component." Unlike the other terms, it strictly concerns the internal component of a given being insofar as it is an internal component. *Arche* and *aitia* may indicate internal or external causes.[7] If a primary principle will constitute an actual part of some being, simultaneously, it will be an element. For example, Thales simultaneously conceived water (*hudor*) as a primordial principle and an element, for water generated all things, and water persists within the beings that arose from it.

The wise person must know the basic elements from which everything was produced. The person who does not possess wisdom may know at best that something took place, while the wise person will seek an answer to the question "why?" (*"dia ti?"*). The Ancient Greeks called things that provide an answer to the question *"dia ti?"* principles, causes, and elements.

The Greeks formulated such questions because reality as we know it is not evident to us. Change exists within reality. And by change beings arise and perish. On the sense level, change, or motion, is what most captures our attention.

Next, we are struck by the composition of particular beings out of such or other components. Without composition, no change would take place and no variety would exist.

The question "*dia ti?*" concerns primary principles, causes, or reasons, and elements or components. Aristotle tells why this is so:

> Because of wonder the people of the present and the first thinkers started to engage in philosophical thought; initially they wondered at the unusual phenomena they encountered daily. Later they slowly began to face more difficult problem, for example phenomena associated with the moon, the sun, and the stars, and the origin of the universe . . . they engaged in philosophical thought to escape ignorance."[8]

The Greeks did not contrive the question "*dia ti?*" The question grew out of ordinary ignorance and wonder, for unusual things happen every day, some things occur only rarely, and the question of the reason for everything, for the whole universe, is always with us. The wise person looks for answers in the primary principles, causes, and elements. This is the search for knowledge for its own sake. "[S]ince people started to engage in philosophical thought in order to free themselves from ignorance, it is clear that they were seeking knowledge itself, not some utility derived from knowledge."[9]

Aristotle called this most basic philosophical knowledge wisdom, first philosophy, and theology. He developed the framework of this domain of knowledge in the fourteenth book that a later tradition called "metaphysics." This is the doctrine of being as being, or the first and basic causes and principles of being, and of the first substance.

Several theories exist about the author and the meaning of the term "metaphysics." The most widely accepted theory is that the author was the last scholarch of the Academy, Andronicus of Rhodes (first-century B.C.). He arranged Aristotle's works so that these fourteen books appeared after (*ta meta*) the works on physics (*ta physika*). So, they were called *ta meta ta physika*. Some scholars hold that the term arose earlier in the third-century B.C.. Hans Reiner maintained that Eudemos of Rhodes was the author of the term. Paul Moraux thought the author was Ariston of Ceos.

The term's meaning is also disputed. Not entirely evident is that the books that are collectively called the metaphysics were located after the works on physics. In the Aristotelian division of the science, first philosophy, as it was called, came after mathematics, not physics.

Some neo-Platonists tried to explain the term by saying that metaphysics is about divine matters: things that are beyond nature. Hence metaphysics as a science concerns what it beyond physics (Simplicius). Yet the term *meta* does not mean something higher and better, since *hyper* means higher and better. So, metaphysics would have been called instead *hyperphysics*. Another explanation: metaphysics is placed after physics because the matters it concerns are the most difficult for people to understand (Alexander of Aphrodisias). The order of knowledge, not the order of things, determines its place. I find this explanation the most convincing.[10]

By nature all people desire to know, and in increasing degrees we perfect this natural inclination. At the start common experience exists. Later, art. Then, science. Finally, the science of the sciences—first philosophy, or metaphysics—which seeks a rational answer to our most sublime questions. We start from impressions that we share with animals, and we rise to wisdom concerning divine matters in theology.

Five

KNOWLEDGE AND OPINION

Our knowledge begins with our senses and rises ever higher until it finishes in discursive thought and intellectual intuition. Along the way, we start to make an important distinction between opinion and science.

Opinion engages different cognitive powers, but essentially differs from science. As we try to describe with precision scientific knowledge in its uniqueness, we must show the difference between science and opinion.

People normally voice many different views, prosaic and sublime, on many topics. The Ancient Greeks were the first to ask whether all opinions are of equal value. Do only some opinions carry weight, and if so, upon what does this depend? In this way, the Greeks arrived at the quite basic distinction between knowledge and opinion, science and supposition. Not all views have the status of being scientific, and not everything is a matter merely of personal opinion.

The Greek philosophers considered the difference between knowledge and opinion with regard to the (1) object to which our knowledge refers; (2) act by which we grasp the object; (3) faculties of knowledge; and (4) knower.

The root of the Greek term "*episteme*" ("science") is the verb "*istemi*" ("*epistemi*"). A very basic verb that enters into the composition of many words and has a wide palette of meanings. The most important meanings for understanding the Greek conception of science are to: (1) "hold fast"; (2) "stand"; and (3) "persist."[1] The term "*doxa*" ("opinion") has three meanings: (1) "to show," or "to indicate"; (2) "to appear" or "to seem"; (3) "phantom," "apparition," or "illusion."[2]

According to Greek etymology, what belongs to science is what we retain in our knowledge. What merely appears to be so, or is an illusion, belongs to opinion.

We have to set retention in opposition to what is variable and illusion in opposition to what is true or of a deeper nature. The variable appears to our senses. A thing's deeper nature is what something external hides.

Science permits us to reach a thing's deeper and more persistent nature. Opinion slides about the surface of the variable outer appearance. Ancient philosophers added many refinements to this distinction. But they preserved this basic meaning of the difference between science and opinion.

From the start, the Ancient Greeks based the difference between opinion and science upon the perceived difference between sense and intellectual knowledge, and the corresponding difference between their respective objects. Our senses perceive the variable. Our intellect perceives the stable. This opposition between sensory and intellectual knowledge, and their objects, had radical philosophical implications in logic, epistemology, and metaphysics.

Parmenides asserted that being is the intellect's object and non-being the object of the senses. For something to be a being is to be self-identical, and to be is not-not-to-be. If something changes, it is and it is not. And so it is not a being as something self-identical. Change indicates loss of identity.

The intellect grasps being and truth because the true is what is. The senses grasp an illusion of being, what is and is not. Since nothing can simultaneously be and not be, the senses show only an illusion of being. The senses lead us into error when we think that we know a being. It is no being at all.

Parmenides made quite a clear distinction between the competence and credibility of knowledge supplied by the senses and the intellect. He characterized sense knowledge by falsehood and error; intellectual knowledge by truth. His distinction between the objects of the two powers had radical consequences: being is the intellect's object; non-being is the object of the senses.

The stability of intellectual knowledge with respect to the concept of being, which occurs in metaphysics, allowed Parmenides to reveal identity, non-contradiction (which Ancient philosophers treated in metaphysics and logic), and truth (which they treated in metaphysics, logic, and epistemology). A negation of identity, non-contradiction, and truth is the instability characteristic of sensory knowledge. Intellectual knowledge is the domain of true knowledge. Sensory knowledge is the domain of misleading opinion.

Plato and his disciple Aristotle made further contributions to the distinction between knowledge and opinion. Plato referred to Parmenides' division between what is stable and variable. He associated science with intellectual knowledge. It takes the form of *noesis* (intuition) or *dianoesis* (discursive thought). Opinion takes the form of *pistis*, belief, or *eikasia* (images).

Science's object consists in ideas we grasp by intuition (*noesis*), or numbers, which we grasp in discursive thought alone and sometimes with discursive thought and mental images (*dianoia*).

Opinion's object is the sensory world, whether we grasp the world in the present by the senses (*pistis*, as our spontaneous confidence in what we see), or in images; for example, in shadows or reflections, or in mental images (*eikasia*).[3]

Plato divided all beings into the spheres of (1) ideas, (2) numbers, and (3) material beings. Science addresses the first two spheres. Opinion addresses the last.

Plato's position on the object of opinion was more moderate than that of Parmenides. Plato maintained that opinion's object is not non-being. It is something between being and non-being; something that does not exist, but is not something absolutely not, for it becomes. Consequently, opinion is not ignorance, but something between knowledge and ignorance.[4]

Hence, in the *Republic*, Socrates says:

Then opinion is not concerned either with being or with not-being? —Not with either. — And can therefore neither be ignorance nor knowledge? —

That seems to be true. — But is opinion to be sought without and beyond either of them, in a greater clearness than knowledge, or in a greater darkness than ignorance? — In neither. — Then I suppose that opinion appears to you to be darker than knowledge, but lighter than ignorance? — Both; and in no small degree. And also to be within and between them? — Yes. — Then you would infer that opinion is intermediate? — No question.

With respect to the act of knowledge, science is essentially intellectual. It is strictly intellectual in the case of *noesis. Dianoia* makes use of images. These images, however, do not belong to the act's essence. *Dianoia* concerns mathematical operations that we aid, for example, by using diagrams. The mathematician knows that diagrams have a schematic function and are mere approximations.

Opinion is not mere sensory knowledge. It does not consist only in sense perceptions. An element of judgment is present. Judgment consists in calming thought while it halts at only one opinion. As Socrates says:

I mean [that thinking is] the conversation which the soul holds with herself in considering of anything. I speak of what I scarcely understand; but the soul when thinking appears to me to be just talking—asking questions of herself and answering them, affirming and denying. And when she has arrived at a decision, either gradually or by a sudden impulse, and has at last agreed, and does not doubt, this is called her opinion.[5]

Even this "halted" opinion, one of many, is not completely permanent. It can change under the influence of emotions, persuasion, or beliefs.[6] True knowledge, however, is unchanging.

Plato restored the status of opinion with respect to human conduct. In the *Meno* he says that reason (or science) is not alone in understanding right action (action in accord with virtue), but true opinion also plays a role.[7] In the *Statesman*, he shows even greater enthusiasm for "true opinion": "I call it divine whenever there arises in souls an opinion that is essentially true and associated with certainty concerning the beautiful, the just and the good, and about every opposite, and I say that it is born in a divine genus."[8]

While Plato's division of reality into ideas and the sensible world underlies his division of knowledge into knowledge and opinion, we cannot ignore some nuances in Plato's views. Sensory knowledge can play a role in intellectual knowledge (*dianoia*). And opinion is no purely sensory kind of knowledge. It implies the participation of the intellect (judgment).

Yet knowledge is only present where stability is. And opinion is present where change exists. Since ideas are stable, only ideas can be objects of science. The sensible world as changing can only be an object of opinion, and opinion will be as variable as its object.

Opinion is important in daily life because it concerns what changes and could happen otherwise. Ideas are stable, while human action involves a kind of change. Since human action should be right and virtuous, we must link opinion with knowledge of the good. In action, knowledge and opinion come in contact with each other. Yet how can we save true knowledge if the theory of ideas has been rejected as unnecessary?

Aristotle rejected Parmenides' theory of being and Plato's theory of ideas. Despite this rejection, he succeeded in saving knowledge and the distinctness of knowledge in contrast with opinion. Knowledge is stable, necessary, and unchanging because even in sensible reality, something stable and necessary exists, serving as the foundation for general concepts. This stable element is substance. Substance has a fourfold causation.[9] With respect to its subject, science is a quality of the subject, a *hexis* (habit) or virtue.

By essence opinion is changeable. So we cannot consider it a virtue or *hexis*. Opinion is a *diathesis*: a disposition easily changed by time's passage, arguments, or feelings.

Opinion is no sense impression or concept. We express it in a judgment that may be true or false. The judgments that occur in virtues such as art, science, prudence, wisdom, and intuition, are always true, if those are real virtues. In contrast, the judgment that occurs in opinion may be true or false.[10] This contrast is not sufficient grounds for dismissing opinion completely as always mistaken and false, as Parmenides did. Opinions are sometimes true.

Because opinion may be true or false, Aristotle introduced the concept of probability, likelihood, or likeness (*endoxon*) as common to all opinion. The probable is not what is like the truth. It is what may be true, and, in equal measure, may be false. Our opinions have this feature of probability.

Truth is about being, and verisimilitude or likelihood is about what appears to be true. A scientific judgment is necessarily true, while the judgment of opinion is necessarily likely.

Likelihood and opinion refer to what is unstable and unnecessary, what could be otherwise, is changing and accidental. Opinion consists in acceptance of an unnecessary premise. Opinion is unstable because that to which it refers is unstable.

In his *Nicomachean Ethics* and his *Topics*, Aristotle expanded his discussion of opinion. He did not restrict opinion to unnecessary and changing matters. The opinion of the majority or of qualified authorities may concern necessary and unchanging things, eternal and contingent things, and possible and impossible things.

Hence, Aristotle said in the *Topics*, "[T]hose opinions are 'generally accepted' which are accepted by every one, the majority, or the philosophers— by all, the majority, or the most notable and illustrious of them."[11]

Opinion includes being and "non-being." Dialectic and rhetoric operate in this field.

Knowledge has an objective and subjective aspect. Objective: the necessary connections in being. Subjective: the permanent ability of the mind capable of grasping these connections.

In its objective aspect, opinion refers to what is unnecessary and accidental. In its subjective aspect it refers to what is perishable and changing disposition in making a judgment about what is unnecessary, or such a disposition in repeating a necessary judgment that we hold by following someone else's ability, not our own.

True knowledge is a fusion of some aspect of reality and the proper habit of mind. Opinion may concern (1) a non-essential aspect; (2) an essential aspect that it treats as if it were non-essential; or (3) an essential aspect following another person's, not our, judgment.[12]

Just as science is the proper type of cognition for grasping necessary connections and the truth, so the likely, what may be true or false, rules dialectic. What is objectively is true because it concerns necessary connections is likely in subjective terms when someone does not possess the proper scientific ability and repeats the thought of someone else. This position is not psychologism. It is a conception of the truth as the ability to see the agreement of judgment with how things are. The truth is more than a property of judgment: a reflection. We may justify an opinion while it is based upon a non-necessary premise. In such a case, the conclusion is merely likely.[13]

While Parmenides so strongly disparaged opinion, opinion won its proper place in culture at the end of the classical era of Greek philosophy. While opinion differs from science, to some degree it absorbs science. Science is not opinion. But an opinion may express a view that, in objective terms, is scientific. A subtle, yet crucial, difference.

Easy to present the general picture of science gaining importance at the cost of opinion, or opinion gaining importance at the cost of science. This position of importance depends upon the culture of knowledge. We may generally disparage science or value it only in view of its role in a worldview or ideology.

In Plato we cannot make a definitive division of science and opinion according to ontological criteria (the World of Ideas and the sense world). And, in Aristotle, we cannot definitively divide science from opinion on the basis of objective necessity or stability, because what is necessary may, in subjective terms, be an object of opinion.

Six

THEORETICAL FEATURES OF THE OBJECT OF SCIENTIFIC COGNITION

In the *Phaedo*, Plato considers knowledge's specific nature. He writes, "Whether it is the blood by which we think, or air, or fire, or none of these, only the brain (*enkephalos*) provides us with the impressions of hearing, seeing, and smelling; then from these memory and opinion supposedly arise, and from memory and opinion, when there is a halt, knowledge (*episteme*) arises in the same way."[1]

Knowledge, therefore, is not a kind of impression (as are vision, hearing, and smelling). It is not a reminiscence of anything or an opinion, because in all these kinds of cognition something is lacking that is present in knowledge. Sense impressions, memory, and opinion are all variable. They change or are liable to change. If knowledge is authentic, it must be stable. Plato explains the etymology of the word "*episteme*" from "*istesis*" (that which stands still); knowledge is what we retain and stands firm in cognition.[2]

In the *Physics* Plato's student Aristotle, wrote of the same feature of knowledge, "[R]eason knows the senses as well by rest and holding firm"; and only that person

> can become rational and fully aware who passes from the natural motion of thought to the stilling of the mind. Therefore children cannot learn or make judgments on the basis of the senses, nor can elderly people; the motion and internal commotion is too great in them.[3]

Rest, retention, and stillness are the characteristic state of someone who has knowledge. When change and commotion occur within someone, that person does not have knowledge.

Knowledge differs from other kinds of cognition because stability is one of its essential characteristics. In other kinds of cognition different kinds of change occur. How can I say that I know something if the object is incessantly changing. At best I could say that I am seeing, hearing, or feeling at this moment, or that something appears now to be so. If my knowledge is not stable, I cannot say that I know.

Upon what does this stability depend? In the texts cited above, Plato and Aristotle indicated that the knowing subject plays a role. Our powers of cognition become unsettled by something and are moved. Then they are rocked. Vision, hearing, and smelling receive a constant and unending stream of impressions.

Only the mind, as both philosophers suggest, appears capable of quiet and thereby of entering into the state of knowledge. The mind can stop in the process of knowledge. Does the stopping of the mind depend on our mental abilities, or does it depend on an object which would be of a different nature from the object of the senses? Are the object of the senses and the object of the intellect different? We must consider the difference in objects. The object possesses features that enable us to grasp it in a state of calm.

Plato completely separated the spheres of sensory and intellectual objects. Sensory objects change, and the sensory powers receive these objects as changing. Intellectual objects are stable. Because they are stable we cannot grasp them by the same powers that we apply to changing objects. We must employ a completely different power: the intellect.

Stable objects (such as the ideas) lie beyond the material world. So the intellect capable of grasping ideas itself is separate from the body. This conception of knowledge met the criteria for the object of scientific knowledge. But at a great price: realism. This kind of scientific knowledge is not a knowledge of the real world that surrounds us.

Is everything in material reality truly changing or variable? Aristotle thought that, while many things change, something is stable whereby things retain their identity and in this way can become objects of science.

Aristotle thought that a concrete being changes and is composite. We may speak of (1) structural compositions within the being and (2) different ways of being: in itself or in something else.

The human being is more than a body, is composed of body and something whereby the body is human, and the human being is a human being. Furthermore, the same human being can be fat at one time, thin at another. He may be sitting at one time and walking at another. He may be sleeping or awake. In all situations he is the same human being.

Something determines this stable identity in a being. Because it is something stable, it can be the basis for understanding and scientific knowledge. Aristotle did not think it necessary to divide the object of scientific cognition from the material world because stable elements exist in the material world. In every being some element must exist whereby the being is the same. An element of identity in every being. It would not exist without stability. For something to be the same, it must constantly be the same. That is, it must retain some stability.

This theme of stability is present in the most metaphysical terms that Aristotle employs. But we may easily overlook this fact in simple translation. Two terms are of special importance: substance and essence.

In Polish and English, our word "substance" ("substancja" in Polish) is derived from the Latin term "substantia" and was used to render the Greek word "ousia." If we were to translate the Latin "substantia" back into Greek, we would have to use the Greek term "hypostasis," which presents the idea of a foundation as something that stands under ("hypo" and "sub" equal under; "stasis" and "stantia" equal standing).

These details of etymology and translation entail semantic shifts that have had serious metaphysical repercussions in philosophy's history that are still with us. The Greek word *"ousia"* is a noun we derive from the verb *"einai,"* "to be." The participle of *"einai* is *"ousia"* ("being," "that which is"). And we derive the noun *ousia* from the participle.

We have no one simple counterpart for the Greek *"ousia."* We can only attempt to render it by an expression such as "that which is being." In Latin the following words occur corresponding to the above Greek expressions: *"esse," "ens,"* or *"essentia." "Essentia,"* not *"substantia,"* should have been the proper Latin counterpart for the Greek *"ousia."* However, the term *"substantia"* ("a foundation") historically supplanted *"essentia."* Thus the dispute about the understanding of being and substance went in a completely different direction from the one Aristotle presented in his metaphysical writings.

Scholars often translate the Greek term *hypokeimenon* into Latin as *"subiectum."* Etymologically, this translates into "that which is thrown under." It is close in meaning to *"hypostasis"* (which would have been the best counterpart for *"substance"*) as that which is like a foundation for properties and accidents, insofar as "that which is being" indicates something in itself that persists without change. If properties exist at times and then do not exist, then "that which is being" is stable and unchanging, is "being." The Greek verb *"einai"* ("to be") does not have a strictly metaphysical dimension. But it primarily designates permanence, stability, and invariability. When we ask about being or *ousia,* we are asking about what is stable in a being.

Medieval writers used the term *"substantia,"* not *"essentia,"* to translate the Greek term *"ousia"* because readers in the Middle Ages were first acquainted with Aristotle through Anicius Manlius Severinus Boethius' translations of Aristotle's logical works. Boethius used the term *substantia* in the logical writings to render the Greek term *"ousia."* In the metaphysical writings, *ousia* appears for the most part translated as *essentia,* and this is the term Boethius used in his translations. The logical translation historically supplanted the metaphysical translation with respect to terminology and philosophy.[4]

Another term quite important in metaphysics is "essence" (Polish—*"istota"*). It corresponds to the Greek phrase *"to ti en einai."* While the word "essence" has an artificial ring to it and is not too transparent in meaning, in the Greek phrase, the verb "to be" appears twice: (1) as the infinitive *"einai"* used as a predicate; (2) as a predication and copula in the form of the verb in the imperfect tense—*"en."* Later commentators attempted to explain this strange construction in many ways. They faced difficulties because Aristotle did not provide any explanation.[5]

No imperfect tense exists in Polish; only the past tense and present tense. The imperfect tense in Greek expresses well the moment of stability and duration, for many of our activities persist in time and connect the past with the present: the present does not artificially and arbitrarily divide from the past.

In Polish we cannot say, as the Greek could, in one verb, that someone fell asleep at eight and is still sleeping now at noon. When we use the present tense, we obscure the reference to the past. In English we might say (in a complex construction that still is ambiguous as to duration in the present), "John has been sleeping since eight o'clock." Although that John still sleeps appears clear, in an expression such as "I have been ill," whether I am still ill is not clear. In Polish we would say "John sleeps already since eight" (*"Jan śpi już od ósmej"*). When we speak of the past, we lose the precise reference to the present ("John fell asleep at eight"—*"Jan zasnął o ósmej"*).

The imperfect verb *"en"* says that something was, not only that something is, but that it was and is. What is it that was and is? It is *einai*: "to be." The formulation *"to ti en einai"* expresses a duration that was and is. The Polish philosopher Roman Ingarden translated the phrase in Polish as *"czym rzecz bywszy jest"*—which we may attempt to express in English as "what the thing, having been, is."

The term "essence" appears as the rendition of the Greek phrase. The Polish term *"istota"* contains the root "ist," which appears in the word *"episteme."* The Polish word does not have as clear a connotation as the Greek expression. So it has an artificial or technical sound to it. In the Latin language, the translator William Moerbeke used a literal word-for-word translation—*"quod quid erat esse,"* for *"to ti en einai."* Otherwise, Latin writers usually used the term *"essentia,"* which was a literal translation of the Greek term *"ousia,"* to render the phrase *"to ti en einai."*

In the Greek terminology, the verb "to be" (*"einai"*) appears in two logically consistent forms in reference to the two key metaphysical expressions, to emphasize the element duration, stability, and invariability. In the Latin terminology and in other European languages that have derived their terminology from Latin, the terms used in place of the Greek do not exactly present the same meaning. The element of stability and invariability is simply absent.

An object of fully rational knowledge cannot exist and then not exist. If rational knowledge will occur, the object must endure. We have no knowledge of non-being. Something must exist upon which the "eye" of our intellect can rest. Stability is then the first feature that the object of theoretical cognition must possess.

The second feature is necessity (*anankaion*). We think of necessity with respect to compositions of being and connections among beings, not with respect to change in beings. As stability is opposed to change, so necessity is opposed to the accidental. Aristotle defines the chief type of necessity: "when something cannot be otherwise than it is."[6] That which can occur otherwise is not necessary. In this sense, necessity is an essential feature of the object of scientific knowledge. If we deny necessity, we deny being in some essential aspect.

Metaphysical necessity would concern the basic causes that constitute being, and the negation of one of these causes (the material, formal, efficient, or final cause) would lead to the denial of being. For the most part, Aristotle calls necessary what is incapable of not being [*Metaphysics* Bk.5, ch. 5, 1015a 33]. We may verify this quasi-definition may in other definitions or descriptions of ontological necessity, the necessity of a thing's nature. [*Metaphysics*, Bk 8, ch.

8, 1050b 11], with the final cause [*Metaphysics*, Bk. 5, ch. 5, 1015a 20], and with the efficient cause [*Metaphysics*, Bk. 5, ch. 5, 1015a 26].[7] Without these causes no being would be. And if a being is, then these causes also are that constitute the being. Scientific knowledge is based on the necessary elements of being.

Necessity plays a role in understanding and scientific proofs. Necessity then takes the form of syllogisms. In a syllogism we arrive at a conclusion that is just as necessary as the premises. Thereby our knowledge of the truth is broadened.

This happens when a syllogism is based on one of the four causes. Aristotle writes: "We think that we know something unconditionally, and not sophistically or accidentally, when we are convinced that we know the cause due to which a thing possesses an attribute, that it is its real cause and cannot be otherwise."[8] When something could be or occur in another fashion, our knowledge loses its scientific value.

Necessity is a theoretical feature of the object of scientific knowledge, because here, as in the case of stability, our reason must have the possibility of immersing itself in an understanding inspection of being in each case. Necessity makes this possible. So it deserves emphasis.

Generality is the last feature of the object of science. Many philosophers have misunderstood this feature, especially since the dispute over universals that led to nominalism, which had serious consequences in philosophy and other domains of knowledge and culture.

A concrete being contains some elements that are liable to change and others that are unnecessary and accidental. It contains matter as potentiality. And matter is indefinite in itself and without its own act. The concrete thing, insofar as it is concrete, cannot be an object of scientific knowledge for nothing exists in it that the reason can conceive in a stable manner. Variation, change, and lack of definition are associated with the material concrete thing insofar as it is concrete. That is what is expressed in the Aristotelian phrase, rendered commonly in Latin—"*individuum ineffabile*"—we cannot adequately express the individual in words.

Yet in every concrete thing something is stable and necessary that may be common to other concrete things within a certain species. An individual named John has features that are not repeated. He shares a common human nature with Adam, Matthew and Eve. The unchanging and necessary aspect of the concrete being provides the foundation for what we express in general concepts.

This generality appears when we are dealing with a material concrete thing, since it is matter that is the reason for accidents, change, and concreteness. A material being is a suitable object for scientific knowledge under the aspect of its stability and necessity, not under its material aspect.

If a being is immaterial, then generality is unnecessary as a feature of the object of knowledge, since the nature of such a being is not multiplied in others. Matter is the reason for multiplication. Generality as a feature of the object of scientific knowledge concerns a definite aspect of material being, the aspect that endures in relation to a knowing subject.

Generality is not a property of the being considered in itself. It is connected with the mode of human knowledge. When we know concrete material things, we must "purify" the concrete thing of what is changing and unnecessary to make the object suitable for intellectual apprehension. In the Latin tradition, the process whereby we arrive at an object adequate to the intellect is called "abstraction."

Generality is a way to grasp the object known, which, as concrete and material, possesses something that the intellect cannot grasp. We cannot say that the material object is unknowable. And we cannot say that it is completely ready for knowledge, since only some aspect of the object is intellectually knowable.

The intellect must arrive at this aspect. When the intellect arrives there, we see that this aspect is also common to other concrete things that belong to the same species. We say in short form that the object of intellectual knowledge is general.

Where we follow Latin with its one term, "*abstractio*," Aristotle used two terms: "*aphairesis*" and "*epagoge*." "Aphairesis" means taking away and separation. Today we understand abstraction as the process whereby we arrive at general concepts; the process of abstraction is a process of generalization. Aristotle saw the problem differently. *Aphairesis* applied only to mathematical objects: to quantities. Quantity is the result of abstraction, but abstraction conceived as cutting off and leaving to the side everything that is not quantity, not as generalization.

How was quantity associated with separation? Quantity is not a being in itself. It is a property of a being. No "2" or "3" exists, but 2 geese or 3 ducks; 100 kilograms cannot exist as such, but 100 kilograms of something can exist. If the mathematician studies quantity considered in itself, he must first "separate" it from the subject to which it belongs and from any other properties. Mathematical objects come from abstraction. We do not take away the mathematical objects, but we take away something else from them.[9]

Aristotle speaks of abstract things only with respect to quantity. Only number occurs in scientific considerations as a quasi-substance: a quasi-being considered in itself. The mathematician studies quantity without its connection with the real subject in which it occurs. Otherwise he or she could not perform many mathematical operations. To add 2 and 3 is easy. But we cannot add 3 ducks to 2 geese. Abstraction primarily concerns quantity separated from the other categories and treated as a being in itself (as a substance).

In Latin we render the Greek "*epagoge*" by "*inductio*." In English, by "induction": the natural process whereby generalities or concepts arise in our minds. The process starts with sense perception and concludes in a general concept. Aristotle wrote:

Thus from sense perceptions arises what we call a memory, and from a repeated memory referring to the same object experience is born, for from many facts of memory a particular experience is born. Again, from experience, that is, from generalities enduring in their wholeness in the soul, of a unity or plurality [of particular sense impressions], which is at the same time one and the same in the plurality, arises that which is the principle of art and knowledge; art is directed at what comes into being, and knowledge is directed at what is.[10]

Knowledge concerns what is, what is unchanging and enduring, whereas art concerns what comes into being. The repeated action of the same object leads the generality to arise. Aristotle observed further:

When one among many details that cannot be differentiated logically is retained, then the first generality arises in the soul, for although the object of sense perception is individual, its content is general, for example, it will be man, but not the man Callias. Again we stop at these first generalities and the process does not cease until indivisible concepts and true generalities are established.[11]

The structure of our cognitive powers is such that generalities arise spontaneously in us as a result of the natural collaboration of the senses and intellect. These generalities concern that which endures. So, they can be a foundation for scientific cognition. That which is merely sensible is concrete and transient. As such, it is not suitable for scientific knowledge.

Generality as a feature of the object of scientific cognition functions in opposition to the concrete and variable, which join only with matter and the senses. The possibility of making generalities is associated with the unique structure of the human intellect.

Aristotle made a hypothetical distinction between an active intellect and a passive intellect. If we are capable of knowing reality by way of concepts, then we must grasp the concrete thing we encounter in such a way that the intellect can interiorize it. In his theory of these two functions of the intellect, Aristotle used the analogy of the relation between sensory knowledge and intellectual knowledge. For the eye to see, the eye and object are not enough. Light and a diaphanous medium must exist.

In the case of intellectual knowledge, the role of the active intellect is to illuminate the concrete thing so that the stable, necessary and general elements are manifest in it. Only then is the passive intellect capable of assimilating or receiving what is suited for intellectual knowledge because it possesses features adequate to the intellect. The object does not enter the intellect. Only its intentional similarity enters.[12]

Generality is not a property of the concrete or of the intellectual cognitive aspect. It is linked with the mode of human knowledge. Generality is a potency, not an act. We know the concrete thing in the present. But we may predicate the content suitable to the intellect of many concrete things that possess this same content. We may predicate a content known in the present of other concrete things. The knowledge of the content in the present is an act; the possibility that it may be predicated of other things in the future is a potency. Generality is a potency. It must be constructed on stable elements in a thing, since these will also be present as stable elements in other concrete things. As Joseph Owens states:

The Aristotelian universal, accordingly, is an individual form considered according to its possibility of being seen in many things, whether those things be KATH'HEN or PROS HEN. As such it plays the leading role in the order of logic. But all that is actual in what the universal denotes is the individual form.

While, from the metaphysical viewpoint, the individual being involved, is the individual act, while nothing universal can be a real individual being, the individual form is the reality implied in the universal. The composite singular finds its actual expression in the form. "In this way and according to these relations," Owens says, "universal, form, and real individual being coalesce as the cause of Being in sensible things. To this extent do universality and individuality coincide, even in sensible forms. The same form, by its very nature, is actually individual and potentially universal."[13]

Stability, necessity, and generality manifest different aspects of the same object of scientific knowledge as theoretical knowledge. In the Aristotelian conception of being, the object that focuses different aspects considered in itself is the form (*eidos*): that whereby the thing is what it is. The form is also (1) the first analogue of substance (*ousia*); (2) the first object of definition (*to ti en einai*); and (3) the central point (or middle term) in syllogistic proof. Without form, substance, being, or scientific knowledge would not exist.

Plato thought that the idea was the culminating point of scientific knowledge. Form has that function in Aristotle. The form is, simultaneously, the guarantee of realism. It guarantees that scientific knowledge is a knowledge of the real world.

Seven

THEORETICAL JUSTIFICATION

Scientific knowledge and opinion respectively have different objects. The object of scientific knowledge must be necessary, stable, and general. The object of opinion is unnecessary, unstable, and concrete. When we express objectively scientific views as opinions, they lose their theoretical character. Scientific knowledge has a unique object and its own mode of rational justification.

We hold many of our views without direct knowledge, but in each case we use some kind of rational justification. Different forms of rational justification exist, including appeals to (1) authority, (2) tradition, or (3) different kinds of causes. Not all forms of rational justification have scientific status. This is evident in Aristotle's discussion of the views of Hesiod.

Aristotle tells us "Hesiod's contemporaries and all the theologians like him provided a kind of explanation (*pithanou*) that could satisfy themselves but did not consider us." They thought that (1) the gods caused everything, (2) principles were of divine origin, and (3) mortality's cause was that some beings were not permitted "to taste nectar and ambrosia."

Given these assumptions, Aristotle concluded that, evidently, they had a private language (*ta onomata*), one only they could understand. Aristotle, however, found what they said about the divine influence of such causes "beyond his ability to understand" (*"hyper hemas eirekasin"*). He reasoned: "If the immortals lived on nectar and ambrosia for the sake of pleasure, then these in no way could have been the cause of their immortality; if they did this because of need, then how could they endure eternally, since they need food?"

Hence, he concluded, that devoting his attention to such mythological subtleties (*mythikos sofizomenon*) was pointless. Instead, he thought , "with those who tried to justify their views (*apodeixeos*)," he should try to discover an answer to the question why, in the end, some things that arose from the same elements have an eternal nature, and others do not.[1]

In Aristotle's response immediately above, we see how mythological or mytho-philosophical explanations cannot satisfy us from a scientific viewpoint. Scientific and philosophical justification must be intersubjectively communicable. It cannot be restricted to a closed circle. And people in general should be able to understand it. It should be free of logical errors and should not employ a sophistic appeal to mythology.

With this conception of scientific and philosophical justification, Aristotle could set aside competing doctrines, especially mythology. We accept these more on the basis of belief and imagination than on the basis of reason.

When he left behind mythological explanation, he paved the way for first philosophy, or metaphysics, and physics, or natural theology. He discredited

Greek, and all, mythology. In so doing, he generated a new feature that differentiated Greek culture from Eastern cultures. He distinguished philosophy and
mythology on the basis of the criteria used to justify their views, not on the basis
of their answers. Philosophically and scientifically, this was of crucial signifi-
cance.

Aristotle did not develop a detailed methodology of metaphysical knowledge. He developed a theory of scientific knowledge for the particular sciences.
In the post-Aristotelian tradition many misunderstandings have arisen as philosophers have tried to apply the methodology of the particular sciences to
metaphysics.

They forgot that the object of metaphysical knowledge is being as being, not
any particular category of being grasped in a particular aspect. The unique object,
being as being, requires a different approach or method than in any of the
particular sciences.

Philosophers in our time have developed a methodology of metaphysical
knowledge. The Polish philosophers Mieczysław Albert Krąpiec and Stanisław
Kamiński, co-founders of the Lublin school of philosophy, made a great contribution to developing this method with their work on this question. In the
introduction to *Z teorii i metodologii metafizyki* [*On the Theory and Methodology
of Metaphysics*], they atate:

> Aristotle himself seemed to follow his own guiding principle, which soon
> after was translated into Latin as *modus sciendi ante scientiam* (the mode of
> knowing before knowledge). In the *Analytics* he presented a theory of
> science with sufficient precision. People in later ages also had this
> impression, since the philosophers of the Middle Ages and the Renaissance
> (mostly within Scholasticism), were convinced that Aristotle's logic really
> did contain the theory of scientific philosophy that those philosophers *de
> facto* practiced. This is also the conviction of many contemporary authors
> (mostly authors of textbooks) who belong to the traditional school of
> philosophy known as Aristotelian-Thomistic philosophy, since when they
> write about their own method of philosophy, as a rule they appeal to the
> Aristotelian conception of deductive science, which would be perfectly
> verified precisely in the philosophy of the traditional school.[2]

How does justification appear in the particular sciences? The problem is that
we do not know everything directly. In many cases we arrive at knowledge by an
indirect route, but our aim is that our indirectly obtained knowledge should be
true. The knowledge of what is simple, or not composite, and of what is always
composite and divided does not completely exhaust the range of scientific
knowledge.

The essential feature of science is proof or demonstration, whereby we may
also know what we could not learn immediately, for whatever reason. As
Aristotle wrote: "[A]nd with the help of proof we acquire knowledge."[3]

"By a proof," Aristotle explained, he understood "a syllogism that produces scientific knowledge, namely one whereby, if we are in its possession, we have this knowledge."[4] A proof is a syllogism. A syllogism is a "statement in which something is presupposed, and something other than the presupposition must result because it was presupposed."

By "because it was presupposed" Aristotle says he understands "that only in view of that fact that it is such as was presupposed, and by that I again understand, that no additional term is needed in order for necessity to arise."[5] Thus a syllogism is composed of premises (that which is presupposed) from which the conclusion necessarily results.[6]

A premise is a statement that affirms or denies something about something. Three terms exist in the syllogistic premises: (1) middle term; (2) the first or major term; and (3) minor or last term. The minor term is the one addressed by the inference, the subject of the conclusion.

In the conclusion, we predicate the major term of the subject (the predicate in the conclusion). The middle term joins the minor term with the major term. We join the predicate to the subject because of the middle term.[7] We join them when there cause exists for the connection: "Thus in all these investigations we ask whether it is a middle term, or what it is, for the middle term is the cause that we seek in all investigations."[8]

What are the causes of which Aristotle is thinking? Aristotle answered: "We think that we possess scientific knowledge when we know the cause; and there are four causes: first, the essence; second, the precedent, which necessarily implies the consequent; third, the efficient cause; and fourth, the final cause. Each of these causes may be the middle term in a proof."[9]

Each of the four Aristotelian causes may be a middle term: (1) the formal cause or essence; (2) the efficient cause; (3) the final cause; and (4), at least in some sense, the material cause. Aristotle mentions the material cause in his second point: "the precedent, which necessarily implies the consequent." Aristotle is predicating the term "matter" in an analogous sense because the premises are that from which the conclusion arises. They are like the matter ("that from which") for the generation of the conclusion. Aristotle is not talking about matter as a pure indeterminate potentiality.[10]

The Aristotelian syllogistic is not based upon purely extensional operation. It involves finding real causes for the fact that one thing belongs or does not belong to another.

All syntactic systems are subordinate to more primary causal relations. Furthermore, since syllogistic relations are necessary, in principle, the concrete thing as a concrete thing cannot be the subject or predicate in the premises and conclusion. Nonetheless, we should not conclude from this that Aristotle had no awareness of syllogisms using particular premises. As Aleksander Achmanow states:

If we speak of the forms of a syllogism's premises and their symbolic forms, we should recall that when Aristotle classified judgments in the work *On Expression* [*Peri Hermeneias*], he spoke of individual judgments, and in the *Prior Analytics*, a work devoted to the theory of syllogism, he does not mention that individual judgments may occur among premises, and he investigates only syllogisms that contain general and particular terms. . ... We may add that Aristotle never introduced any symbols for individual judgments, and where this is necessary, in the role of the individual term, he uses a proper name, for example: "Socrates is white." We should not think, however, that Aristotle had no knowledge of syllogisms with particular premises." [11]

Aristotle's theory of proof was the first attempt to show philosophically the principles that govern scientific justifications. The greatness of his theory is that it was a pioneering effort, and in it the concept of scientific knowledge as scientific, distinct from opinion and myth, started to crystallize.

This theory of proof was strongly rooted in the Aristotelian theory of being (metaphysics). Syllogistic was not a separate domain, like a syllogistic game of a few rules and principles. So we cannot consider it only at the level of syntax.

The fundamental purpose of justification was to acquire true knowledge in accord with the criteria characteristic of theoretical knowledge. It should start from necessary, general and invariable premises and arrive at conclusions with the same features. This would be a scientific syllogism, and it is part of the essence of a scientific syllogism that it should lead to true knowledge.

Merely following some rules of inference is not enough for a syllogism to be scientific. Aristotle speaks directly of this issue:

If knowledge is such as we have shown it, then the premises of demonstrative knowledge must be true, primary, immediate, better known and prior (to the conclusion) and must be the cause of the conclusion. In this way, the principles also meet the condition that they should be proper for the proven fact. A syllogism may also arise without these proper premises, where as a proof may not; for it does not produce scientific knowledge." [12]

Just as no essences are in non-existing or merely possible things, so no proof fails to provide true knowledge. A reasoning that was only a purely syntactic operation could be called a syllogism, not a proof. Only a real being possesses an essence, and a proof leads to true knowledge.

A name to which nothing corresponds in reality possesses a meaning, not an essence. As Aristotle says in answer to the question, "How shall we by definition prove essential nature?", "He who knows what human—or any other—nature is must know also that man exists; for no one knows the nature of what does not exist—one can know the meaning of the phrase or name 'goat-stag' but not what the essential nature of a goat-stag is." [13]

A syllogism that does not lead to true knowledge is no proof. Aristotle's position is quite clear about the issue of scientific proof and shows the fundamental role that realism played in his thinking on this matter.[14]

Eight

THEORETICAL PROOF

Truth nourishes theoretical life (*theoria*). The true philosopher loves beholding truth.[1] Truth, however, is not easy to acquire, and not everyone can reach it. Plato wrote in one of his letters: "For while it might be thought that excellence in courage and speed and strength might belong to different men, everyone would agree that surpassing excellence in truth, justice, generosity and the outward exhibition of all these virtues naturally belongs to those who profess to hold them in honor."[2] (Worth noting is Plato's use of the term "excellence," not "value" in his text.)

We often count truth, or, more strictly, love of truth, among the greatest virtues, along side magnanimity and justice. To know the truth is difficult. And most people do not consider that to acquire the truth we must go through an entire field to find our way.[3]

In Plato's vision of reality, we find truth chiefly in an unchanging and immortal World of Ideas, not in that which changes and perishes.[4] The World of Ideas nourishes the soul and is the source of authentic happiness. Plato said, "[S]ince the divine mind is nourished by thought and pure knowledge, therefore the mind of every soul that would seek to achieve what is worthy, when it looks at the Being for a long time rejoices, and beholding the truth it grows and fills with happiness until after the circuit of the circle the circular motion leads the soul to the Being."[5]

Plato maintained that contemplation (*theoria*) of the highest ideas and Being itself makes true happiness (that of the rational soul) complete. He intermingled truth with the highest Good and Beauty.[6] Thus, the truth is, simultaneously, the aim of our desire and object of our contemplation.

Plato thought that human beings and the gods treat the truth as the highest Good.[7] A human being who wants to be perfect should, therefore, follow the truth.[8] We can know the highest truth, which shows what each thing truly is, only with the help of the reason.[9] "The truth is the property of a thought formulated as a judgment about what is, that it is" (*to ta onta dokazein aletheuein dokei soi einai*).[10]

Theoretical truth possesses an epistemic aspect for Plato, a property of a thought or judgment that asserts that something is. But this "is" (*einai*) must concern what truly is. Here the ontological aspect appears—an idea is that which truly is. An idea is a true being. Theoretical truth is based on the conformity of a judgment with an idea.

In the Aristotelian theory of being, form, or essence, occupies the place of ideas as the internal constitutive component of every being. The double aspect of truth as epistemic and metaphysical also appears in Aristotle's meditations.

Aristotle writes: "To be means also to be truth, and not-to-be nor not-to-be-true is a falsity both in affirmative and negative statements. For example, it is true to say that Socrates *is* a musician, that he *is* not-pale, and it is false that the diagonal of a square is equal in measure to the side."[11]

The word "is" appears as a copula in a sentence to indicate composition in being. Affirming agreement of a composition in a judgment with a composition in a being means our statement is true.

Likewise, when we state that no composition exists, and, no such composition is in a being, this agreement is also a sign of a true judgment. For we assert that no composition exists. And no composition exists. Falsity occurs when we assert that a composition exists when none does, or we assert that no composition exists when some does.

In another passage Aristotle wrote that as truth and falsity respectively, being and non-being refer to composition and division. Simultaneously, they constitute two members of an opposition. Truth consists in asserting the real composition and denying the division. Falsity consists in denying one or the other assertion.

Aristotle added that we base a judgment's truth on that which exists in reality. Truth does not exist because we think something is so. It exists because something really is as we say it is. The real composition is the cause of a true judgment asserting that the composition is true.

Truth and falsity depend on composition and division in things. This means that we are in a state of truth when we regard the really divided as divided, and the really combined as combined. We are in a state of error, or falsity, when we judge otherwise than things are.

Aristotle asked us to consider the conditions under which we say that truth and falsity exist. His answer was that a person is not white because we hold he is white. The reverse is the case. If a person is white, and we say so, we speak the truth.[12]

Things may be permanently, or at some time, united or divided. In the second case, a judgment cannot always be true or false. "In reference to what may be such or otherwise," Aristotle maintained, "the same opinion and the same statement may be true and false: at one time it may be true, and at another time false. As for things that cannot be otherwise, there is not truth about them at one time and falsity at another, but the same statements concerning them are either always true or always false."[13]

From the theoretical viewpoint, the truth that science, or philosophy, seeks is that truth that concerns what is always true.

Another problem arises about truth. What is truth and falsity with respect to what is not composite? Where can the truth of knowledge not be based on affirming or denying a connection or division in things?

Aristotle's answer was that something composite exists when it is combined and does not exist when it is divided, or separated. A non-composite being, a simple substance, does not come to be by being combined. It does not cease to be by being divided. From this situation Aristotle concludes that truth and falsity will be different when they refer to uncombined and combined beings.

In the case of non-combined beings, truth consists in intellectual contact (agreement) with what is and in a statement, an expression, that verbally signifies what something is. This is not the same as a verbal expression that signifies a composition in something that is. And ignorance (falsity) regarding simple substances consists in lack of agreement (lack of contact). It involves no verbal expression at all because, just as our natural faculties cannot be mistaken with respect to some proper object, except perhaps in an accidental sense, so our intellect cannot be mistaken with respect to non-composite, or simple, substances. If we are in contact with them, we know them. Not to know them, to be in a condition of falsity regarding them, is not to be in contact with them.

All such substances exist in act, not in potency. Otherwise they would come into being and be liable to decomposition. Simple substances do not come into being and are incapable of destruction because they do not come to be through generation.

Consequently, Aristotle maintained that we cannot be mistaken about the fact that something is a something (an essence) and an act. We know it or we do not.[14]

Truth is a property of affirmation or denial. Both statements are composite and concern the attribution of one thing to another (composition or connection) or non-attribution (division). Truth exists when we affirm that one thing belongs to another and, in reality, it belongs to the other as we affirm it, or we deny attribution and, in reality, the one thing does not belong to the other. Falsity exists when we affirm combination when, in reality, division exists, or when we deny a connection and, in reality, a composition exists.

The way we predicate truth is consequent upon the way things exist. Some beings, however, are such that they are always composed and cannot be divided. For example, a human being's rational soul cannot be separated from its sentient nature. Some beings are such that they may be combined and separated. For example, a subject capable of possessing differences that can be combined in it, like sleeping and black in John. Some beings are separated and cannot be combined, like contrary or contradictory opposites. In the case where we affirm or deny combination of a composite subject, truth expresses unity. In the case where we deny combination of a composite subject, truth expresses pluralization or otherness.[15]

With respect to what is simple or non-composite, truth cannot consist in uniting or pluralizing. In this case, truth consists in contact with what is simple and involves affirmation thereof. Falsity consists in lack of contact. In the case of what is simple, we know or do not know.[16]

In science, the truth includes states of composition or division that always take place. They include knowledge of what is simple. And this knowledge is, in a sense, higher than the truth. Compositions and divisions that may and may not be are the object of *doxa* (opinion) and are considered in rhetoric and poetry.[17]

Theoretical truth is always so and may refer to compositions and divisions, while the state above truth refers to what is not composite. The non-composite and our knowledge of the non-composite belongs in an eminent degree to the order of *theoria*, because, since it is simple, it is identical with itself, stable, and unchanging.

The Greek conception of truth as theoretical is objectively directed at reality. The Greeks were primarily interested in the stable and unchanging aspect in reality.

Plato could not see anything stable in material reality. In the name of science he arrived at his theory of ideas. Aristotle linked science with material reality by his discovery of the form and essence. Ontological truth concerns reality itself to the extent that reality is stable. We may refer stability to compositions in being and to what is simple. In terms of cognition, theoretical truth with respect to composite things consists in expressing in a judgment the unchanging fact of composition or division (that it is, or is not). With respect to simple things, truth is an infallible act of knowledge whose character is higher than the truth.

Nine

TYPES OF SCIENCE

According to Diogenes Laertius, Plato divided the sciences (*tes epistemes*) into practical (*praktikon*), productive (*poietikon*), and theoretical (*theoretikon*). Diogenes explained this division thus: The productive sciences, for example, the science of building, are those whose products are visible; the practical sciences, such as politics, or playing the flute or lyre, do something, but make no material products. The theoretical sciences produce nothing and perform no actions. Their concern is investigation (*theorei*). The geometer investigates the relations among lines; the musicologist studies sounds, and the astronomer investigates the stars and the cosmos.[1]

We find this division of the sciences nowhere in Plato's extant texts. But, since this division appears many times in Aristotle's works, this division has been traditionally associated with Aristotle, not Plato.

Each science's aim is the criterion of its division. The poetic sciences' aim is production (*poiesis*—production, making). The practical sciences aim at action (*praxis*—action, operation). The theoretical sciences aim at knowledge alone (*theoria*).[2]

The product in *poiesis* differs from mere activity. The finished house differs from the act of building. The Greeks would later call poetic sciences useful sciences. In *praxis*, the activity is the end and no product exists apart from the activity. Other European languages have assimilated the term "practical" in another sense. It has come to mean "useful," which is not implied in the Greek term *poetikos*.

Also, *praxis* does not include art as presented in Diogenes' examples (playing the flute and lyre). It is restricted to the sphere of our moral life: *praxis* is behavior or conduct. The Ancient Greeks thought of religion as part of πραξις [*praxis*—practical action], and associated it with the virtue of justice, as the honor due to the gods. Christian tradition would later distinguish religion as a distinct cultural domain.

The end in the theoretical sciences, where no product exists, is the activity of knowledge. Only the action performed in the acquisition of knowledge exists. While a house differs from the activity of building, a person is just precisely in just conduct and not apart from such conduct.

Likewise, we understand in knowing, and not apart from knowing. We cannot understand if we do not know. We cannot see if we do not look. The division of the sciences into productive (or poetic), practical, and theoretical logically includes the whole of our rational life.

The productive sciences include the different industries and arts. According to the peripatetic, or Aristotelian, tradition, we divide the practical sciences into ethics, economics and politics. Ethics studies moral behavior in personal life. Economics, in the area of family life. Politics in social life.[3] The theoretical sciences include physics, mathematics, and first philosophy (first called theology, and later called metaphysics).

The criteria for the division of the theoretical sciences have been a topic of controversy throughout history. This division is most like Plato's distinction among the three spheres of reality: (1) ideas (studied by dialectic or first philosophy), (2) numbers (studied by mathematics); and (3) the sensible world (studied by physics).

Aristotle rejected the sphere of ideas and numbers, and accepted the existence of the sensible world and the superlunary world. Even if, in place of ideas, in Aristotle's thinking, first philosophy were to study the divine, the presence of mathematics in Plato was not intelligible, and no sphere corresponded to mathematics in Aristotle's teaching.

The Scholastics tried to rid this division of its Platonic baggage. They looked to three orders of abstraction for the criteria of division. The three orders of abstraction moved from physical, through mathematical, to metaphysical.

While many Scholastic manuals perpetuated this approach, it was not completely in agreement with Aristotle's views. Aristotle associated abstraction only with mathematical beings. He did not speak of abstraction with respect to physics or first philosophy (theology or metaphysics). One way or another, the Aristotelian division of the sciences into theoretical, practical, and poetic (or productive), along with their respective subdivisions, was the first attempt to look at science and human culture in an integral way.

Later in the Roman period, when theoretical philosophy gave way to more practical attitudes, writers made this threefold division of the sciences into a threefold division of arts or skills. Hence, the Roman rherotician Quintilian maintained:

There are other arts. Some consist in observation, that is, in knowing and estimating things. An example is astronomy, which requires no practical action but is content with an understanding of the things it studies. Such arts are called *theoretike*. Other arts consist in action. Their end is action. An art of this kind is perfected in the act itself and leaves nothing after the act of performance. This kind is called *praktike*, and dance is an example. Other arts are consummated in their result, that is, they attain their end in a completed concrete work that comes under visual perception.[4]

Discovery of *theoria* as knowledge for the sake of knowledge bore fruit in the appearance of many particular sciences. Development of a methodology of scientific knowledge for these sciences (in Aristotle's *Posterior Analytics*) was its crowning glory.

Simultaneously, a new domain of knowledge arose to take first place among the sciences on account of its (1) object, (2) relation to the other sciences, and (3) importance in knowledge and life: first philosophy, later called metaphysics.

Aristotle put the crown on the Greek movement of *theoria*. He thought that wisdom is the summit of knowledge and being for human beings. And he maintained that wisdom consists in encompassing in our understanding all being as being under the aspect of its principles and causes.

Most valuable to philosophy and science is that, during the classical period, the Greeks succeeded in saving science as a unique domain of knowledge in the face of different difficulties. Plato saved science from the sophists, but at the cost of separating science from the real world. By showing that scientific knowledge could still exist and have the real world as its object, Plato's student, Aristotle, enabled the Greeks to connect science with the reality. The introduction of science, especially of realistic science, was crucial to culture. Thereby, Western culture started to differ essentially from the Eastern cultures. The battle for the place of science in culture and the battle for the realism of science would be an indication of the frictions between the West and the East.

Part Two

THE HELLENISTIC DEFORMATION OF THEORETICAL KNOWLEDGE

Pandora by Hugh Mcdonald

Ten

KNOWLEDGE OR PHILOSOPHY FOR THE SAKE OF *PRAXIS*

Although Athens eventually lost its political independence, philosophy was still able to follow the path marked by Greek thought. During the Hellenistic period, after Greece lost its political freedom and Greek culture spread in the lands conquered by Alexander the Great, philosophy lost its theoretical character. Part of the reason for this change was that many things in Aristotle's *Metaphysics* were unclear and needed to be expressed more precisely, but they were not. As a result, Aristotle's *Metaphysics* became forgotten.

Another reason for this change relates to a loss of Aristotle's works. According to a legend passed on by Strabo and Plutarch, the manuscripts of Aristotle's esoteric works (works intended for a specialized circle of readers) were entrusted by Aristotle's student Theophrastus to his fellow student, Neleus. Neleus' heirs were uneducated and hid the manuscripts in a cellar because they were afraid of the greed of the Kings of Pergamon.

In the first century B.C., to the peripatetic philosopher Apellicon of Teos bought Aristotle's works. Lucius Cornelius Sulla seized his library during the war and took the books to Rome. Tyrannion the grammarian bought them, and Andronicus of Rhodes bought the books from Tyrannion. Andronicus also published Aristotle's esoteric works (around 60 B.C.), including the *Metaphysics*. Aristotle intended his exoteric works for a broader reading public. And they were in constant circulation in the Ancient world. But these works have not survived to our time.[1]

Other problems, for example in ethics and physics, needed resolution in view of metaphysics. This did not happen.

The schools of the great masters (Plato's Academy, Aristotle's Lyceum) and newer schools (Epicurus' garden, the Stoa) took philosophy in a practical direction and, at best, preserved a few of the particular sciences. *Theoria* as *theory* no longer had any meaning. While the founders of the Cynic school, Epicureanism, and Stoicism addressed the problems presented by Socrates, they forgot the strictly theoretical solutions of Plato and Aristotle.

Socrates had ordered his philosophical investigations to the question of human happiness. He maintained that we could not achieve happiness apart from morality and virtue. Later, Aristotle's analyses would go beyond the problem of morality and into the realm of metaphysics. Still later, philosophers would return to the Socratic questions, but would forget metaphysics.

Stoicism represented this attitude with its syncretic approach. Stoicism has no *theoria* in the strict sense, and no metaphysics. Metaphysics reappeared in the Hellenistic milieu within neo-Platonism. For different reasons, neo-Platonism moved beyond *theoria*. *Theoria* would not return to its proper place in scientific culture until the Christian thought of the late Middle Ages.

Skepticism was a third school of Hellenistic philosophy, beside Stoicism and neo-Platonism. Pyrrho of Elis (376–286 B.C.) is the generally recognized founder of Skepticism. Philosophers have most often presented this complicated and influential school in the light of philosophical arguments intended to show the uncertainty of human judgments. This is a more important aspect of the influence of Skepticism because it undermined the entire anthropology of Greek philosophy. Simultaneously, it denied the essential meaning of the first sentence of Aristotle's *Metaphysics*, which states that the desire for knowledge is innate in human beings and culminates in the βιος θεωρητικος (*bios theoretikos*—the life of contemplative knowledge) and wisdom.

The Skeptics taught that the desire for knowledge only brings people unhappiness. Therefore, we should become indifferent to knowledge. Under their influence, *bios theoritikos* ceased to be the apex of culture.[2]

1. Stoic Modifications to Philosophy

Zeno, from the Phoenician city of Kition, or Citium (333 or 332–262 B.C.), founded the Stoa. When he was young he had followed his father in business, the proverbial occupation of the Phoenicians. According to one legend, Zeno found himself in Athens by accident as the result of a shipwreck during a voyage of commerce. Another legend relates that Zeno's father, Minaseas, brought Xenophon's *Memorabilia* from Athens and gave it his son as a gift. Zeno developed such an ardent love of philosophy that he left his island and went to Athens. According to a third legend, Zeno asked a seer what kind of life was in store for him. She answered that he will speak with the dead. Zeno interpreted this to me that he would be devoted to reading the Ancient writers and that he should go to Athens to do this.[3]

Whatever the case about what caused Zeno to come to Athens, once he got there he acquired a taste for philosophy and became a student of Crates, whose views came from Socrates by way of Diogenes and Antisthenes.[4] Zeno's other teacher was Stilpo, a student of the Thrasymachos of Corinth. Thrasymachos had been a student of Euclid, founder of the Megarian school. And Euclid had been a student of Socrates.[5]

In this way, Zeno encountered the moralistic trend in Greek philosophy that treated ethics as the most important domain. Zeno remained faithful to this approach and developed it into the philosophical teaching that would later be called Stoicism, from the word *stoa* (porch).[6] Other philosophical movements of the time (such as Cynicism, Epicureanism, and Skepticism) influenced this teaching. But the Stoics carried on a polemic with these other teachings. Unfortunately, all Zeno's writings, and many of the writings of his successors, have perished.

Philosophers customarily distinguish three periods in the history of Stoicism:

(1) The late Stoa—Zeno of Citium (Kition), Cleanthes of Assos (342–232 B.C.), Chryssipus of Soloi (277–204 B.C.); (2) the middle Stoa— Panetius of Rhodes (185 or 180–110 B.C.), Poseidonius of Apamea (140/130–59/40 B.C.); (3) the early Stoa—Seneca (4 BC–65 A.D.), Musonius Rufus (around 30 A.D. to around 100), Epictetus (50–130 A.D.), and Marcus Aurelius (121–180 A.D.).[7]

The chief representatives of old Stoa were not Athenians, or even Greeks. They came from the neighboring Semite peoples (the Phoenicians). Besides Zeno, we should mention Chryssipus, Antipater, Archimedos of Tarsus, Heracles of Carthage, and Boethius of Sidon. In the next generation of Stoics, were Diogenes and Apollodorus, both from Babylon.

In the second-century B.C., Stoicism started to dominate the Roman elite and would later have a great influence upon Christianity. Stoicism was no uniform philosophical school. Since we no longer have any proper textual sources on the old Stoa, and only fragments on the middle Stoa, to make an exact analysis of its doctrines is difficult. Yet the basic features of Stoicism remained without change.

2. The Stoic Understanding of Philosophy

Seneca wrote in his letter to Lucilius:

Sapientia perfectum bonum est mentis humanae; philosophia sapientiae amor est et adfectatio: haec eo tendit quo illa pervenit. Philosophia unde dicta sit, apparet; ipso enim nomine fatetur quod amet. Sapientiam quidam ita finierunt ut dicerent divinorum et humanorum scientiam: quidam ita: sapientia est nosse divina et humana et horum causas. Supervacua mihi haec videtur adiectio, quia causae divinorum humanorumque pars divinorum sunt. Philosophiam quoque fuerunt qui aliter atque aliter finirent: alii studium aliam virtutis esse dixerunt, alii studium corrigendae mentis; a quibusdam dicta est adpetitio rationis.

(Wisdom is the highest perfection of the human soul. Philosophy is the love of wisdom and the persistent striving after wisdom. Philosophy aims where wisdom has already arrived. The origin of the term "philosophy" is clear. By its name, philosophy professes what it loves. Some have defined wisdom in such a way as to call it the knowledge of divine and human matters. Others again say that wisdom is a knowledge of divine and human matters as well as their causes. This addition seems to me superfluous, since the causes of divine and human matters are a part of divine matters. Also, other people have defined philosophy in other ways. Some called it the study of virtue, and others called it the study of perfecting the mind, and yet others called it the desire for reason. . . .Philosophy is the science of virtue, but the science that is acquired by way of virtue.)[8]

Seneca listed several conceptions of philosophy: (1) a science about divine and human matters (sometimes including their causes), (2) the study of virtue, and (3) the correction of mind. We should not let the term "divine" mislead us. It does not refer to gods beyond the world. These gods exist in the world as pirnciples of movement in things. Hence, they pertain to physics.

Seneca's distinctions match the Stoic division of philosophy into three areas: (1) physics (the philosophy of nature), (2) ethics (moral philosophy), and (3) logic. These divisions are complementary. Yet the Stoics did not agree among themselves as to which should hold the first place.

Zeno thought that ethics is the most important. Chryssipus thought that logic should hold first place. Ultimately the Stoics thought that "the single most important thing is to know how to live well and happily": "*aei to idion ergon ektelousa, ten euzoian*" (Clement of Alexandria).[9] The result was that most Stoics, including Zeno, recognized ethics as philosophy's first domain.

In this way, philosophy ceased being theoretical wisdom and became practical wisdom. Hence, in *Metamorfozy pojęcia filozofii* [*Metamorphoses of the Concept of Philosophy*], Julius Domański tells us that the common opinion of the philosophers from Plato to Plotinus was that "to be truly a philosopher" we must know more more than how to spend our lives: we must know "how to live in accord with this knowledge. The life of the philosopher, his conduct and personality, are thus a completion of the whole and integral conception of philosophy."[10]

The basis of this wisdom was not so much maxims or experience in life. It was different sciences, especially the divisions of philosophy mentioned above. While the Stoics continued to have a practical inclination, they proclaimed a cult of scientific wisdom.

3. The Divisions of Philosophy

The Stoics introduced a division of philosophy that eventually would be more influential than Aristotle's division. They thought that logic, physics, and ethics composed philosophy.[11] Today, philosophers generally regard this division as most representative of the Ancient Greeks. More than Aristotle's division, it dominated Hellenistic and Roman thought, and has also been quite influential in the modern era. As Ernst Cassirer says:

The division of philosophy into three main fields—logic, physics, and ethics—was already complete in antiquity and has continued to be firmly maintained every since, virtually unaltered. Immanuel Kant still recognized this threefold division as valid, declaring that it conforms perfectly to the nature of things and permits of no improvement.[12]

The Stoics illustrated their division with different metaphors. Some compared philosophy to a garden: logic was the hedge; physics, the tree; ethics, the fruit was. Some compared philosophy to an egg: logic was the shell; physics, the albumen; ethics, the yolk. Others compared philosophy to a living animal: logic was the skin; physics, the flesh and bones; ethics, the spirit.[13]

Sextus Empiricus ascribed this division to the Platonic school, in particular to Xenocrates.[14] Plato made no precise designation of philosophy's divisions. In his *Republic* he said that some sciences are practical (*praktiken*) and some are theoretical (*gnostiken*).[15] Sextus vindicated Xenocrates by claiming that this threefold division was present in Plato in a potential sense (*dynamei*).[16]

We do not know the precise origin of philosophy's threefold division into (1) ethics, (2) physics, and (3) logic. And do we do not know how Xenocrates "lost" some of the sciences along the way and changed the basis of division. If he was referring to Socrates, Socrates did not mention physics. If he was referring to Plato, Plato spoke of the theoretical sciences. In fact, when Zeno arrived in Athens he encountered a division of philosophy that was a version of the circle of Socrates and Plato. Xenocrates included this division in his own teaching. This division was later universally acknowledged as Stoic.

When we compare the Stoic with the Aristotelian division, we see that, in the Stoic division real impoverishment exists in the domains of knowledge. Aristotle first divided the sciences into theoretical, practical, and poetic in view of the aim of knowledge. Each domain differed according to its object. Aristotle described the status of logic as ancillary, as an aid.

The Stoics abandoned these criteria and divided philosophy in a completely different way. Of all the sciences that fall under *theoria*, only physics remains. Of those belonging to *praxis*, only ethics remains, while economics and political science vanished. The domain of *poiesis* vanished completely. Logic, which for Aristotle was part of the *organon*, a tool of knowledge, became a science under the Stoics. These changes signaled an essential change in the conception of philosophy and science. We will now consider the origin of these changes.

4. The Leading Science: Ethics or Physics?

Zeno and most of the Stoics regarded ethics as philosophy's most important domain. This was clearly part of the Socratic heritage.

But Socrates was part of a broader current in Greek thought in which the problem of *eudaimonia* (happiness) appeared along with the Ionian *theoria*. The Greek philosophers pondered about where we could find true human happiness. Socrates considered the many answers to the question and concluded that we find true happiness is the fruit of moral virtues. Ethics studied the virtues. Therefore, if the most important thing for us is to acquire true happiness, the most important domain must be ethics.

Plato and Aristotle regarded acts of theoretical knowledge our highest acts. They held that we find *eudaimonia* in acts of knowledge, while Socrates regarded moral acts as the location of *eudaimonia*.

The Stoics followed Socrates in this matter. For this reason, ethics would have to take the place of metaphysics.

The Stoics re-introduced into science a domain that Socrates did not hold to be of value: physics. The Stoics generally regarded physics as the second science after ethics, as necessary for the sake of ethics. But some Stoics, such as Panetius or Poseidonius, thought that physics was the most important science.[17]

The problem is at the core of Stoic philosophy. Socrates regarded physics as unnecessary, since we understood the good by a dialectical and maieutic analysis of the conception of the good. The Stoics took another approach: they regarded the good as what was in harmony with nature; to be happy is to follow the commands of nature, since nature knows best what is good for a human being. Seneca maintained, *"Incertum est et inaequabile quidquid ars tradit: ex aequo venit quod natura distribuit."* ("What art teaches is uncertain and unequal: what nature gives comes from what is equal.")[18]

An important consequence of this approach is that we must know nature to follow nature. Physics seeks to know nature. So physics became important. Without physics ethics could not exist. The Stoics differed from the Cynics and Epicureans because the Stoics attached great weight to the natural knowledge of nature as a necessary condition for the moral good. To be happy we must follow nature; to follow nature we must know it; and, so, we must be physicists.

Stoics such as the above-mentioned Panetius and Poseidonius thought that physics is the leading science. However, they were not, thereby, following the division of the sciences into theoretical and practical. For them science was still practical. They thought that if the wise person knows nature as a whole, he will necessarily act well.[19]

In this way the Stoics arrived at moral intellectualism by their analysis of nature, whereas Socrates arrived at the same position by his analysis of the good. Whether physics or ethics is the first science, both sciences have a practical purpose. Good action results from the understanding of nature and bad action results from ignorance of nature.

5. Logic as a Science

Socrates used logic, especially dialectic, but not physics, to defend ethics. He wanted to prove that we could reject on rational grounds the relativism and moral subjectivism that the Sophists proclaimed. In his logical defense of ethics Socrates discovered induction and definition. Induction provided a method for discovering the traits common to several things, and definition provided a way to describe these common features. With these methods he discovered that the virtues that correspond to concepts such as justice and fortitude were stable, not changing and relative.

The Stoics advanced the case for recognizing logic as a science and not a mere tool. They presented two main arguments defending logic as a science. (1) They maintained that a skill is only a tool if it serves several domains distinct from it. For example, the craft of the metal-smith is a tool in architecture since it does not serve architecture alone but other arts as well, but surgery is not an instrument of medicine since medicine is the only art that uses surgery. The Stoics thought that logic is applied only in philosophy and so it is not a tool. (2) They said that logic has its own proper object: acts of reasoning—*logoi*, which was not shared by physics or ethics. So logic is a science (Ammonius).[20] Logic, like physics, is theoretical science. Logic considers the practical end of obtaining the good and happiness.

Émile Bréhier claimed that, according to the Stoics:

> [I]t is the same faculty of reason that connects consequences and precedents in dialectics, connects all causes in the area of nature, and establishes perfect agreement among acts in the area of conduct. It is impossible for the good man not to be a physicist and a dialectician at the same time: it is impossible to achieve rationality separately in these three domains, as, for example, the reason cannot achieve a general knowledge of the process of the events that follow one another in the cosmos, unless at the same time it achieves rationality in its conduct.[21]

The Stoics reasoned that dialectics implies physics because a necessary connection between propositions implies a necessary connection between events with regard to their destination. Dialectics also implies ethics because ethics is the virtue that contains the virtues. Virtue is a question of judgment. Likewise any discourse on physics or ethics implies logic because the discourse is a logical statement. Ethics implies physics because as Chryssipus said the difference between good and evil came from Zeus and from universal Nature. Finally, physics implies ethics, insofar as the end-purpose of a rational nature is to know the world and the goods, and where the perception of rationality in events implies the rationalization of moral conduct.[22]

If my end is happiness, and if I find happiness only in morality, then, to act rightly, I must act in harmony with nature—I must know physics. This, in turn, requires me to have a right understanding of nature, which is the task of dialectics.

The physicist must also be a dialectician and an ethicist, just as the dialectician must be an ethicist and physicist. The physicist must be (1) a dialectician because physicists need right reasoning; and (2) an ethicist to know why to seek knowledge of nature. Dialecticians must be physicists to know (1) what they know (nature) and (2) why (ethics).

In this way these three domains of philosophy integrally connect. The Stoics also often stated that, when we speak of one philosophical domain we should speak also of the other two, no matter which one is our starting point.[23] However in view of philosophy's practical end as a whole, we must recognize ethics as the leading domain.

6. Consequences of the Stoic Understanding of Philosophy

Because the Stoics made ethics first philosophy, ethics had characteristics of theoretical knowledge. One such feature was ethical intellectualism, which earlier had been Socrates' position. Moral evil comes from defective knowledge of the good (as Socrates taught) or of nature (as the Stoics taught). Since knowing is objectively the highest human act, and since ethics was put in first position in philosophy, the theoretical element was primary in ethics. The theoretical element was prudence in so-called "popular" ethics, and wisdom in the ideal morality of the elite, a morality accessible only to a few.

Hircilius introduced this kind of distinction. He attributed a completely sovereign good (*telos*) to elite morality, and an inferior end (*hypotelis*) to popular morality.[24]

If these types of cognition are correct, good action must necessarily flow from prudence and wisdom. An action's start is of no concern because it will occur in any case. Right thinking (logic) and right understanding of nature (physics) become the true problem.

Moral intellectualism reaches further and becomes especially evident in the Stoic theory of the cardinal moral virtues. The other cardinal virtues besides prudence are fortitude, temperance, and justice. These last three virtues are species of prudence. Fortitude is right judgment concerning what we must endure. Temperance is right judgment concerning what we must choose. Justice is right judgment concerning our use of things.[25]

We no longer understand virtue as Aristotle understood it, as an acquired skill belonging to different faculties. (For example, temperance is an acquired skill in the area of desire, fortitude is an acquired skill regarding feelings of anger, and justice concerns the will.) For the Stoics virtue becomes exclusively an acquired skill of reason: prudence.

The ability to make a right judgment and find the golden mean is one thing (this belongs to reason). The ability to act that resides in the disposition of a particular faculty (reason, emotions, will) is something else. The will, not reason, wants. The emotions, not reason, feel. Reason is aware of what the will and the emotions do and can guide them to some extent. But reason does not perform their acts. The moral intellectualism of the Stoics brought about important changes in the conception of morality.

Sapientialization of ethics was another consequence of Stoic influence. The Stoic's moral ideal was the sage with an attitude of *adiaphora* (a state of external inactivity) toward the external world. The sage has no influence upon the external world and the happiness or unhappiness in it. Wisdom has influence only over our internal mental state. The wise man concentrates upon himself and his inner state. This leads to the state of *apatheia*—apathy or indifference—which is necessary for wisdom and happiness.[26] The ideal of the wise man who looks within was foreign to the spirit of Greek culture and closer to Eastern ideals.

When ethics supplanted philosophy as a purely theoretical discipline, the result was moral intellectualism and then aretological intellectualism (*arete* is the Greek term for virtue). Finally, ethics focused on wisdom, and so came the sapientialization of ethics.

The Stoics often emphasized that a strict connection exists among the philosophical disciplines. Ethics needs physics and logic, physics needs ethics and logic, and logic needs ethics and physics. Also, despite their practical Easternizing of philosophy, the Stoics regarded the question of which field was primary as open. Diogenes Laertius wrote, "some begin from logic, with physics coming next and ethics at the end. This is what Zeno, Chryssipus, Archedemos, and others did. Diogenes and Ptolemy started from ethics. Apollodorus put ethics in second place. Panetius and Poseidonius started from physics." Others thought that these disciplines are so closely connected that they must be presented simultaneously. [27]

The source of these divergent views among the Stoics and their basic difficulty was that, when *theoria* as *theoria* is absent, we cannot establish the objective order of the sciences or their real relations to one another. Except for *theoria*, no chief science exists that can serve as a reference point, and no science has a proper distance from itself. As reason reflects upon its own act, so *theoria* makes possible a reflection upon the first theoretical science. In the science that studies all reality or being as being, metaphysics, we can see an objective differentiation of beings and the hierarchy among beings. This provides us with a really objective hierarchy of the sciences. If this science is absent the criteria for establishing the order of the sciences must be somewhat arbitrary.

In the question of which science is the primary we must also consider Aristotle's distinction between orders of being (*quoad se*) [in itself] and of knowing (*quoad nos*) [in relation to us]. The orders of knowledge (and learning) and being are distinct. We see first the less essential aspects of a thing. Then, by learning and study, we come to know the more important aspects.

What is first for us in the order of knowing is not first in the order of being. That which is primary in the order of knowing, teaching, and learning is not identical to the importance, in the order of being, of a particular field in relation to other fields.

Condition of the knowing subject influences cognition as an operation of a that knowing subject. As St. Thomas Aquinas says, "*Modus cognoscendi rem aliquam, est secundum conditionem cognoscentis.*" ("The way one knows something is according to the condition of the one knowing.")[28] We start knowledge from the senses (from what is concrete and material), and later arrive at what is essential and general.

If Stoic philosophy has a practical, not theoretical, end, ethics must be the most important science. But this does not mean that ethics will be the first science to be taught. In fact, ethics will be presented last. If the end of philosophy is theoretical, theoretical physics should be the first science. If the end is critical, logic should be the first science. In the absence of *theoria* as *theoria* these questions are difficult to resolve. The theoretical ends run parallel to the objective order, which we can later modify as we take new points of view. Without *theoria* the hierarchy of sciences becomes a matter of convention and depends upon criteria drawn from outside science.

The Stoics sometimes presented physics as first philosophy. Many authors who discuss the physics of the Stoics list its parts as cosmology, psychology, the theory of natural law, metaphysics, and theology or theodicy. The Stoics did not divide physics in this way. But, in their physics, they dealt with questions belonging to many different domains. For example, Marcia L. Colish mentions cosmology, psychology, natural law, and theodicy. Michel Spanent refers to cosmology, theology, and metaphysics. And Georges Rodier writes that, until the end of the Middle Ages, beyond questions of nature, physics studied questions about man, the whole cosmos, and its causes.[29]

In its own fashion Stoic physics continued the thought of the Ancient Greek physicists, the pre-Socratics, who were the first Greek philosophers who investigated the principles and causes of the whole cosmos.[30] When the Stoics put physics in the position of first philosophy, they produced a philosophical teaching based upon materialistic, *a priori*, assumptions. This teaching would lead to a kind of pantheism.

In his *Physics* Aristotle analyzed nature and encountered problems beyond the competence of the physicist as such. So he saw the need for a first philosophy higher than physics. Hence, he said, "The investigation of whether being is one and motionless does not belong among the questions of the philosophy of nature"; and, he added, "The accurate determination of the first principle with respect to form, whether there is one form or many, what form is and what forms are, is a matter of 'first philosophy', and we leave the question to that domain."[31]

Any attempt to resolve questions about the first causes (*aitia*) in physics must sooner or later lead to some form of pantheism, whether we are dealing with the Platonic deification of the World Soul or with the influence of Semitic mythology. Bréhier is one of several writers who support the view that Greek philosophy incorporated elements of Semitic thought by way of Zeno and other Stoics, especially the Semitic idea of God as omnipotent and the director of human destiny.[32]

Physics does not correctly resolve strictly metaphysical problems. Plato had consigned "higher" matters to dialectics. And Aristotle saw the need for a science beyond physics, which he called "first philosophy," wisdom, and theology. Later thinkers would call this subject metaphysics. When the Stoics returned to the pre-Socratic idea of physics as first philosophy, they were choosing a road that would lead to pantheism.

The Stoics considered logic a science that included rhetoric, dialectic, and some elements of epistemology.[33] When they treated logic as a science the Stoics were in opposition to the Peripatetics, who regarded logic merely as an instrument of the sciences, not as a science in itself. I presented above the first argument of the Stoics for logic as an independent science. But their arguments are easy to refute because we use logic in philosophy, other sciences, and in all human discourse.

The Stoics' second argument has a philosophical background. The Stoics rejected the ideas of Plato and the concepts of Aristotle. This had important consequences. Plato introduced ideas and Aristotle introduced abstract concepts to provide science with a fitting object. What is general could provide the stability that constantly changing concrete things lack. What is general is a more fitting object for the intellect than is matter. And the sciences are primarily suited for grasping matter.

In Plato the form or idea is the soul of dialectic. In Aristotle substance, form, or essence as the chief manifestation of being and knowledge is the soul of logic. The Stoic rejected generalities and focused instead on concrete things. Despite this they wanted to uphold the scientific character of knowledge.

Stoic rejection of generalities was connected with two presuppositions: (1) We cannot speak of intellectual knowledge that is solely directed at generalities; and if we know only concrete things insofar as they are concrete, our knowledge will be exclusively sensual. (2) If, despite everything, this knowledge is to be scientific, it must possess the two remaining features that set scientific knowledge apart from non-scientific knowledge: stability and necessity. Can science endure if we reject all generalities? The Stoics thought that it could endure provided we accept some presuppositions from outside of logic.

Determinism and pantheism were the ultimate philosophical or metaphysical foundations of the Stoic conception of logic and science. If we are to reject ideas and forms, the connections that logical propositions express can be necessary only if determinism governs everything and nothing is fortuitous. Judgments are necessary when they concern necessary events.

Why are events necessary? To find the answer we must look to Stoic physics and its pantheistic presuppositions. Absolute determinism is possible only if God, *Logos*, or Spirit envelops, permeates, and has power over the entire cosmos. Concretistic logic is impossible without determinism. And determinism is impossible without pantheism.

Pantheism is hidden below the relations of cause and effect that Aristotle analyzed. To see apparently accidental events as necessary, we must assume a different point of view. Stoic determinism is not identical with the determinism of scientific theories in our time. Stoic determinism did not concern a linear or sequential connection of events. It was ultimately based upon ordering all events to the *Logos* or Spirit as the center. Events that seem accidental because they are not connected are ultimately connected with the *Logos* by sympathy and are thereby not accidental but necessary. This is not necessarily in the Aristotelian sense. As Émile Bréhier says:

> [*L*]*es événements du monde sont liés les uns avec les autres; parce que qu'ils dépendent tous du destin; ils ne se produisent pas les uns les autres, mais ils sont tous produits par une cause unique, identique avec les lois du*

monde. Mais pour atteindre cette raison universelle et réelle de la liaison des événements nous sommes forcés de sortir de la dialectique.

(The world's events are tied to one another because they all depend upon fate. They do not just happen. A unique cause, identical with the world's laws, produces them. But to reach this universal and real reason for the connection of events, we are forced to leave dialectic.)[34]

Stoic determinism subordinated all events to one primordial cause and provided a foundation for a logic of concrete things that, without this presupposition, could not meet the requirements of scientific knowledge.

In this vision of reality the major burden of logical formulations shifts from the subject, or substance, to the judgment as a whole. In the Stoic conception of reality as purely material no place exists for substance. Events have a central position because they imply a connection with the *Logos*.

A judgment renders the fact of an event. So, from a philosophical viewpoint, the judgment becomes the most important part of a logical statement. The judgment, not the subject of a proposition, becomes the basic point of reference. For this reason the so-called logic of names, wherein the subject or substance has the leading role, becomes replaced by the logic of propositions, where events are central. The new logic of the Stoics was not the result of an evolution or revolution within logic. It developed at a deeper level outside of logic, at the frontiers of physics, metaphysics, and theology.

The Stoics rejected concepts and causal connections between beings, with the exception of the *Logos* as the single cause that operates centrally. As a result, their philosophy no longer sought to explain reality in terms of its causes. It broke with the etiological tradition of Greek philosophy (the tradition of explanation in terms of cause). This tradition was typical of the Greeks and unique to their civilization. Stoic philosophy became a form of hermeneutics, a method of using signs to reveal a hidden reality, and, in this case, the hidden reality was the divine *Logos*.

The connections among beings are not causal. They are signs that something more profound is the reason for an event. According to Gerard Verbeke, "[*A*]*u lieu de partir des effets pour découvrir les causes, les philosophes du Portique s'appuient plutôt sur des signes en vue de dévoiler des réalités cachées.*" ("Instead of starting from effects to discover the causes, the Philosophers of the Porch, on the whole, relied upon signs with a view to unveiling hidden realities.")[35]

In this perspective the philosopher becomes a prophet, exegete, and seer. He becomes the interpreter of the signs he sees. But he is not what the Greek tradition would call a philosopher, for a philosopher was one who knows causes. Verbeke adds, "[*S*]*elon les Stoïciens il [le philosophe] est le médecin de cet organisme vivant qu'est le monde; il est aussi une sorte de prophète, un devin, un exégète, un interprète des signes qu'il observe.*" ("According to the followers of

Stoicism the philosopher is a physician of the living organism that is the world; he is also a kind of prophet, divine, exegete, an interpreter of signs that he observes.")

Verbeke also tells us that Stobaeus was one among several writers who thought that only the wise man is a seer, because he is able to understand signs that come from the gods and from demons, and then connect them with human life.[36] At best, the Stoic philosopher is a hermeneuticist.[37]

The Stoics had no place for metaphysics. But they devoted much attention to, and wrote many works on, the theory of language and signs. As Verbeke says, "*Pour les Stoïciens la notion de signe n'est pas un concept secondaire et accessoire, elle est au centre de leur logique, en rapport étroit avec leur façon de penser, leur manière d'interpréter le monde et la conduite humaine.*" ("For the followers of Stoicism, the notion of sign is not a secondary and accessory concept; it is at the center of their logic, in strict conformity with their way of thinking, their manner of interpreting the world and human conduct.")[38] Their interest in language and signs was no matter of scientific curiosity. It was based upon a crypto-metaphysics and a religious outlook.

This is the core of Stoicism: the most important discipline for science is not logic, or the theory of language. It is hermeneutics as a means of reaching a deeper divine reality.

Quite likely, interest in Stoic logic in the nineteenth and twentieth centuries, although garbed in terms of science and the cult of pure science, shares the same hermeneutic subtext and pantheistic presuppositions. The cult of mysticism is not far from the cult of logic, later joined in these centuries with the cult of mathematics.

Eleven

NEO-PLATONISM:
AN END BEYOND KNOWLEDGE

Plato's Academy continued without interruption from Plato's time to the sixth-century A.D.. Historians of Philosophy generally divide the development of Plato's Academy into three periods: (1) The "Old Academy" in which Plato's close disciples set the tone; (2) the "Middle Academy," which was active during the early Roman Empire; and (3) the "New Academy" that Plotinus started and endured until Emperor Justinian dissolved the school in 529 A.D..

Despite, in name, belonging to the same school, the Academicians were not always careful about fidelity to Plato's views or about developing them. For long periods other schools, such as Stoicism or Skepticism, influenced the members of the Academy. In the third-century A.D. a philosopher appeared who did much to return the Academy to Platonic sources, and, simultaneously to leave his own mark on the Platonic teaching: Plotinus, the chief representative of the school known as neo-Platonism.[1]

Plotinus was born in 204 A.D. in Lycopolis, today's Asyut, in Upper Egypt. He was educated in the spirit of Greek culture and became interested in philosophy at the age of 28. In Alexandria the Platonist Ammonius Saccas impressed him. Plotinus studied with Ammonius for eleven years. In 243, Plotinus was one of a group of scientists who accompanied the Roman Emperor Gordian III on an expedition against the Persians in hopes of learning from the Persian and Indian sages. The expedition ended when Gordian was murdered in Mesopotamia, and Plotinus escaped with difficulty.

Plotinus thereafter never took part in any more Eastern escapades. From Mesopotomia he went to Rome. He did not return to Alexandria. He lectured in Rome for a decade before his students persuaded him to put his thoughts into writing.

After another ten years Porphyry became his student. Porphyry wrote a biography of Plotinus. He also collected, arranged, and published the works of his master. Plotinus' works are known as the *Enneads* because the works are organized into nine parts.[2]

In his works, Plotinus refers to Plato, Aristotle, and the Stoics. But he also presents an original version of Platonism. So philosophers have described his views as "neo-Platonism."

Plotinus provides a metaphysical interpretation of Plato. His teaching as a whole is more consistent and organized than the original Platonism.

Aristotle's philosophy also plays a large role in Plotinus' thought. In the first-century B.C. the edition of Aristotle's work Andronicus of Rhodes had created reacquainted the philosophical world with Aristotle. The philosophy of the Stoics also became widely known. At the time a strong tendency existed to find points of agreement among different teachings, especially Plato's and Aristotle's. This contributed to a new philosophical synthesis.

Also, many traces of Eastern thought exist in Plotinus' work. So, many that historians in the nineteenth and twentieth century thought Plotinus' philosophy, while Greek in external appearance, actually represented an influx of the East into Western culture. Hints of Persian religion and Hindu thought are especially present. We do not find any explicit mention of these Eastern influences in the *Enneads*. And we can still regard Plotinus' teaching as a rational continuation of Greek philosophy that provided solutions to the problems that concerned Greek philosophers.

Not all neo-Platonists were Greeks. Plotinus himself came from Egypt, Iamblichus and Damascius were Syrians; Porphyry, a Phoenician; Proclus was from Lycia. When Emperor Justinian closed the Academy in Athens, the "seven Greek philosophers" who fled to Syria included one Syrian, two Phoenicians, one from Gaza, and three from Asia Minor (from Cilicia, Frygia and Lydia).[3]

In neo-Platonism philosophy was no longer the highest domain of culture, which meant that *theoria* as such would not be so highly esteemed. The most important task for the sage was to unite with the supreme deity whom Plotinus calls "The One." This unification takes place at a higher level than cognition, and so mysticism is higher than philosophy. Plotinus still regarded philosophy as the path to unification. But many of his followers were inclined to abandon philosophy for something that resembled Egyptian theurgy. With unification with The One as the new supreme end of philosophy, philosophy declined and the order of a culture based on *theoria* was upset.

The Stoics subordinated philosophy to morality, and the neo-Platonists subordinated philosophy to mysticism. In both cases a change occurred in the understanding of philosophy. And, in both cases, philosophy's position in culture became reduced.

1. From Philosophy to Mysticism in Plotinus

Plotinus constructed his teaching according to a plan based upon emanation. The One stood at the summit of all things. The One was the source from which the form of a being, a hypostasis called Intellect (*Nous*), emanated. It came forth like rays shine forth from the Sun. From Intellect emanated the next hypostasis: the World Soul (*Psyche*), the principle from which the world emanates and which animates the world as if it were one organism. Matter was the last thing to emanate. The procession ended with matter.

The motion of the universe started from, and returned to, The One. In the first phase (*prohodos*) the successive hypostases emerged until matter came to be. The second phase (*epistrophe*) started from matter and returned to The One by way of the hypostases.

My concern at this point is with The One, the first hypostasis, *Nous*, and a human being's position in the process. Plato's philosophy did not clearly define the status of the highest idea (of the Good) in relation to the other ideas, the Demiurge, or reality in general. In Aristotle's philosophy the highest being was a self-thinking thought (*noesis noeseos noesis*).

Plotinus drew together some elements from his predecessors and introduced an innovation. The Intellect, or *Nous*, knows itself (as Aristotle taught). But it also knows all ideas, starting with the new, basic idea of being. The ideas do not exist in a mystical *pleroma*. They exist in the Divine Intellect. That Intellect cannot be the highest principle because a difference exists in it between the Intellect as the subject who knows and the ideas as the objects known.

Something that contains differences cannot be absolutely first because it is not absolutely simple. In holding this Plotinus consistently followed Plato and Aristotle. The Good (The One) is situated at the summit of the cosmos and is absolutely simple. It is The Good because everything emanates from it. It is the primordial One because it is one and absolutely simple.

We cannot say that The One is Intellect or a being because it is higher than Intellect and being. Intellect and being are lower than The One and originate from it. Consequently, many difficulties exist in defining The One and a human being's relation to it. We may speak of The One only in negative terms, saying only what it is not, because we draw all positive descriptions from this world, which lies below The One. The language of superlatives does not provide any new information. It only adds the suffix "above-." Greek philosophy thus arrived at a point where it had to move beyond itself and aim at something higher, which, by its nature, could not be intellectually grasped.

Like the Stoics, Plotinus regarded philosophy as a way of life intended to lead human beings to happiness. By the same token, if philosophy is a requirement for true happiness, that we should go beyond philosophy appears surprising. Can we reach The One? If so, how?

To find the answer, we must consider our position in the cosmos. Unlike his predecessors, Plotinus did not think that the human soul was cast out of a higher realm upon the Earth for some supposed fault or any other reason. He thought that the highest part of the human soul remains in the higher world. As a result, we start from a different point as we seek to realize our highest human aspirations. The lower part of the soul sinks to the Earth. Thereby, we forget our true calling. In this life we must activate our higher part and then the most important perspective, unification with The One, will open before us.

According to Plotinus, art, morality, and philosophy, in succession, prepare us and activate the highest part of the soul. We must become as similar as possible to The One so as to unite with it. The lower something is in the hierarchy of being the further it has emanated from The One, and the more it differs from its original source. The closer something is to The One, the more similar it is to it. The higher part of our soul remains at the level of the first hypostasis (Intellect or *Nous*). It is closest to The One and, as such, most like The One. But it only needs art, morality, and philosophy to open it toward The One. Plotinus said,

"[T]o see the sun, one must become sun-like." Art, morality, and philosophy help to make our soul similar to what we want to behold. The One is above Intellect, being (the ideas), and the human soul. How can our soul aspire to ascend to The One? If The One is invisible and does not know itself, how does it appear? To answer these questions we encounter the mystical element in Plotinus' theory. We need ecstasy (*ekstasis*), a movement beyond ourselves. Plotinus writes: "This is no longer contemplation (*theamata*), but another type of seeing, a going beyond oneself (*ekstasis*) a simplification (*aplosis*), self-renunciation, a desire for contact, a holding."[4] Plotinus further explains that these descriptions are approximations (*mimemata*). They cannot adequately express the soul's state united with The One because this is ineffable and unknowable in a cognitive and contemplative sense.[5]

Porphyry said that, from the time he first met Plotinus, Plotinus had four such ecstasies, while Porphyry only had one when he was 68 years old. Some historians of philosophy (for example, Edward Zeller, Émile Bréhier) have thought that Plotinus' mysticism exhibits a great similarity to Eastern mysticism. This could suggest an inspiration from the East, which is possible since trade routes from India and elsewhere intersected in Alexandria. From the philosophical viewpoint, however, Plotinus' mysticism is philosophically consistent. The source of its consistency is a logical development of Plato's and Aristotle's views. Plotinus took their thought a step further as philosophical thought. Moreover, Plotinus dissociated himself from Gnosticism and theurgy, the lines of thought most evidently the property of the East.[6]

With this background, a problem arises from the viewpoint of the conception of philosophy, when philosophy becomes a road to mysticism. How does this affect the conception of philosophy? Does it not open the door for a complete rejection of philosophy?

2. Theurgy in Place of Philosophy (Iamblichus)

Plotinus' teaching about union with The One by ecstasy became the thought that integrated the entire neo-Platonic movement. Simultaneously, some philosophers had some reservations that led to the proposal of another way than that Plotinus had mentioned. The way of Plotinus was open only for a few because few philosophers existed. And, without philosophy, Plotinus maintained that we cannot achieve ecstasy.

This problem was important to the people of the time. Because of Stoicism, and later, Christianity, people increasingly started to conceive of happiness as something open to everyone, not only to chosen individuals or specialists. Unfortunately, Plotinus' theory greatly narrowed the number of the blessed. The conception of the Syrian Platonist Iamblichus (240–325) appeared upon the canvas of Plotinus' followers, as a counter-position to Christianity. This conception became a major support for the re-paganization of the Roman Empire proposed by Emperor Julian.[7]

Iamblichus, a student of Porphyry, presented another way to union with The One, open to all, in which philosophy was no longer necessary. Theurgy took the place of philosophy and provided a more universal and effective opportunity for union with The One. Iamblichus eliminated philosophy from the Plotinian *epistrophe* and, in its place, introduced Eastern esoteric knowledge. This elimination of philosophy was intentional. Plotinus' philosophical teaching about the image of the cosmos and the end of human life make this procedure possible. In this respect, we may consider Plotinus' teaching philosophical suicide that Iamblichus carried to its logical conclusion.

Iamblichus' critique of philosophy as a way to union with The One was unsparing in his criticism of the Greeks for their instability and their constant search for innovations.[8] This critique was necessary and subsequently praised the wisdom of the Egyptians, as Plato had done. Iamblichus had found in Plato's texts some fragments that lent credence to the generally-held opinion at that time that Plato's doctrine did not deviate from the sacred tradition of the Egyptians, Chaldeans, and Assyrians. In these texts, Plato had said that his writings were only an introduction (*propaideia*) to deeper mysteries. And he did not conceal the influence of Egyptian "wisdom" on his views.[9]

Iamblichus took advantage of such support a reductionistic attack against philosophy. Thus he could write to Porphyry: "You should understand that since the Egyptians were the first who were allowed to share in the life of the gods, the gods are satisfied when they are invoked after the Egyptian model."[10] Iamblichus continued: "The barbarians, since they are attached to their customs, constantly use the same words. Therefore they are loved by the gods, and the invocations directed to them please them. No man is allowed to modify these prayers in any way."[11]

Iamblichus maintained that Greek philosophy is not necessary to enable us unite with, or share in, the life of the gods. This union, or sharing, requires Ancient rituals and Egyptian prayers. At the moment he said this, Iamblichus parted with philosophy, and replaced philosophy with theurgy. While Porphyry still regarded theurgy as a possible, vicarious, means for those who do not know philosophy to ascend to The One, Iamblichus put theurgy in the first, superior, position.[12]

The word "theurgy" appeared for the first time in an occult work called the *Chaldean Oracles*, written by two Julians (father and son), during the second half of the second-century A.D., probably during the reign of Marcus Aurelius.[13] This work played a crucial role in the "Easternization" of Greek philosophy.

Because the *Chaldean Oracles* acquired the status of a holy book, it played a significant role in the battle against Christianity. Porphyry was the first philosopher to cite the work. He wrote commentaries on it in a neo-Platonic spirit. Porphyry's commentaries were necessary because the *Chaldean Oracles* had been written before Plotinus and so contained an earlier version of neo-Platonism close to that of Numenius. Numenius spoke of two Intelligences, not one. As a result, later neo-Platonists in the Athenian School regarded the *Oracles* as having greater authority than Plato himself. So it remained until the Academy was closed.[14]

"*Theourgos*" was a neologism probably coined by the younger Julian, while "*Theurgika*" was the title of a work he wrote as part of the *Chaldean Oracles*. As Hans Lewy writes: "The noun [*THEOURGOS*] is constructed after the model of *THEOLOGOS*: just as theologians are *HOI TA THEIA LEGONTES*, theurgists are *TA THEIA ERGADZOMENOI*."[15] Thus theologians merely speak of the gods, while theurgists perform operations that can actively unite us with the gods.

Iamblichus wrote a work entitled *Master Abammon's Response to Porphyry's Letter to Aneb*. During the Renaissance, Marsilio Ficino composed the new title by which it later became famous: *De mysteriis*.

In this work Iamblichus explained the difference between philosophy and theurgy. He maintained that the philosopher thinks about God or the gods, while the theurgist reaches the god by correctly performing some rituals and magical operations (*erga*). The uninitiated observer cannot understand the actions of the theurgist. But the gods understand what the actions signify and respond in proportion to the knowledge the theurgist possesses.[16]

Iamblichus still taught that philosophy can help someone reach the gods by purifying his soul. But he claimed that (1) we reach the highest state by theurgy and (2) therurgy is superior to knowledge. He said that human intelligence can see the Ideas but not their source. Contact with the source requires a higher organ of understanding called "*anthos nou*," or "the flower of the intellect." This state is higher than knowledge and resembles the divine madness or *mania* of which Plato spoke.[17] A person needs the proper disposition for such a state, what Iamblichus calls a "theurgic" disposition.[18]

Theurgy used magical methods to make divine power physically present in material objects such as statues, stones, or even human beings.[19] The adept could use such objects to go into a trance to raise himself higher and higher (*anagoge*).[20]

Proclus tried to provide a rational explanation for magic with the theory of cosmic sympathy that neo-Platonists and Stoics recognized.[21] In this theory each part of the universe reflects every other part and the material world as a whole is a reflection of invisible divine forces. Because of this interconnection, an action performed with the proper skill upon one element could have an effect upon a completely separate and distant element.

The neo-Platonists thought they could draw powers from the Heavenly spheres by performing specific operations upon different stones, plants, and animals in which the power of the Sun or the stars was reflected.[22] Plotinus did not think that magic could affect the world of the Ideas, and especially The One. Plotinus had spoken of a race of "divine people" who wanted to climb above what was pleasurable and beautiful to the senses. He had asked: "What is this other place and how it is accessible?"

His answer was that some people are born with a lover's nature and authentically philosophic temper. Such people experience pain of love toward immaterial beauty. Not held by material loveliness, they seek refuge from their pain in psychic beauties, in things like "virtue, knowledge, institutions, law and custom."

From there they rise to the source of this loveliness, the Soul. Then to the Intelligence, until they reach, The One, the Principle:

> whose beauty is self-springing: this attained, there is an end to the pain inassuageable before . . . we must look still inward beyond the Intellectual, which, from our point of approach, stands before the Supreme Beginning, in whose forecourt, as it were, it announces in its own being the entire content of the Good, that prior of all, locked in unity, of which this is the expression already touched by multiplicity.[23]

But Iamblichus thought that magic, as a way of domination, could lead even to union with The One. The way to The Good and the Primordial One runs through the vestibule of Mind and philosophy. Particular symbols and rituals could elevate us to that which is supreme.[24] The theurgist calls forth divine forces without which such an ascension would be impossible.[25]

The question of artificial and natural prophecy (*mantika*) was popular in the Stoic and neo-Platonic tradition. Artificial prophecy is the ability to interpret signs. The astrologer, who reads the stars, possesses it. Natural prophecy is the ability to read the first causes contained in the divine mind while the seer is asleep or in an ecstasy.[26] Iamblichus gave prophecy a crucial role because it is an effect of union with the transcendent Intelligence and, thereby, natural prophecy became divine prophecy.[27]

Iamblichus wrote: "Theurgy has a twofold character. It is a ritual performed by a man whereby we preserve the natural order in the universe, and it is also strengthened by divine symbols (*synthemata*), and thereby we can ascend higher toward union with the gods and harmoniously join their order. This second aspect may be rightly described as 'taking on the form of the gods.'"[28]

With this power the theurgist eventually becomes the Demiurge and brings order to the cosmos.[29] In this way the Platonic philosopher and statesman become the neo-Platonic theurge and Demiurge. And philosophy provides a rationale for its own elimination and replacement by magic.

Neo-Platonism was certainly attractive. But it was also dangerous to philosophy and culture in general. It influenced Western culture in many ways. Christianity inadvertently played a crucial role in this influence as it assimilated some of the saner elements of neo-Platonism.

While neo-Platonism heavily influenced Christian thought, especially through the works of Pseudo-Dionysius the Areopagite, Christianity also stood in the way of a dangerous Easternization of philosophical thought. The neo-Platonic version of philosophy decreasingly appeared to be philosophy and increasingly absorbed the practices of the Eastern magic.

This tendency made philosophy quite attractive to many people with its promise of mystery and glamour, but the cost was that it was no longer philosophy. Many Medieval thinkers made clear methodological distinctions between philosophy and theology, reason and faith, and the natural and supernatural. This enabled them, especially St. Thomas Aquinas, to restore philosophy in the original Greek sense while not cutting off rational human thought from the order of Revelation.

Neo-Platonism by itself would lead to the fall of philosophy. This was due to the domination of philosophers from Eastern civilizations and to the idea that the end of human life is something higher than knowledge. Christianity regarded the beatific vision as our end and, thereby, philosophy and theology could stay in their proper places.

Part Three

PHILOSOPHY AND THEOLOGY
IN RELATION TO REVELATION

Pope John Paul II, Mieczysław Albert Krąpiec, Piotr Jaroszyński

Twelve

AVERSION TO PAGAN AUTHORITY

Does revelation that rests on divine authority need any help from pagan wisdom? Some early Christian authorities thought than Christianity had no need of pagan culture. They thought that, insofar as it was pagan, that culture posed a threat to faith, could lead people away from God, from the true religion, and to heresy. They maintained that pagan philosophy, especially, was dangerous because it strove to supplant the *Gospel* and divine wisdom.

One consequence of this aversion to Ancient culture was that many Christians rejected philosophy, which, they thought, should have no place in a Christian culture. This way of thinking has always been present in Christianity, to this day. In this chapter I will examine the sources of this attitude and consider the arguments of some early apologists. I will end the chapter by examining a letter by Pope Gregory IX to the professors of the University of Paris wherein he warns against the introduction of secular sciences and philosophy in education.

Tatian, a Greek apologist from Assyria (b. 120. A.D.), was one of the first Christian authors to disparage Greco-Roman culture. He was acquainted with, and disliked, Greek philosophy. Around 150 A.D., Tatian converted to Christianity.[1] He followed some pre-Christian and non-Christian Hellenic scholars who thought that the Greeks did not discover or invent the sciences and arts, and that they had borrowed their knowledge from other peoples such as the Babylonians and Egyptians. Tatian repeated the opinion of Joseph Flavius that the Greeks had added only their own errors to what is found best in the *Bible*. He thought that Greek culture, including philosophy, is of no value and even harmful.

The writings of another apologist, Tertullian (155–230 A.D.), from Carthage, provide a developed critique of Greek culture and philosophy. Tertullian asked metaphorically, "Who among the poets or sophists has not drunk from the fountain of the prophets?" The assumption in this question was that the *Sacred Scriptures* were the ultimate inspiration for the accomplishments of Greek culture. According to Tertullian, the philosophers had distorted the message of Sacred Scripture.[2] God and nature do not lie. But the pagan books can lead our minds astray. Tertullian also condemned some aspects of Roman public life, such as sports, games, and exhibitions. He regarded them as in discord with nature, reason, and God.[3]

Tertullian raised several objections to the philosophers. The philosophers had distorted the message of the revelation of the *New Testament* by attaching it to a philosophical teaching. They changed the simplicity and certainty of religious truth into something complicated and full of doubt. Whatever in these philosophies agrees with revealed truth, they ascribe to some other source, or they give it a different meaning. The philosophers also attach great weight to the study of nature while they have too little fear of God, which is the beginning of wisdom. They approach the study of the *Sacred Scriptures* with preconceptions, and their conclusions are identical with these preconceptions.

Tertullian also remarked that many philosophical schools, with different teaching views, exist. Even when they agree in some things they still have differences. All this would suggest that the philosophers are far from the truth, that their schools are really sects, and that their views are heresies. According to Tertullian, philosophy wanders about in a mist for it has lost sight of true causes and principles. And, even if it has some grasp of the truth, it is only part of the truth and has been distorted by philosophical treatment. The Carthaginian apologist cites the words of St. Paul (*Corinthians*, 2:8) after a visit to Athens, that we should beware of "subtle words and philosophy."

Tertullian thought that all heresies were rooted in "subtle words and philosophy" and that philosophy was worldly wisdom cut off from God. Philosophers such as Heraclitus and Zeno identified God and matter. The philosophers (1) often dabbled in magic and astrology, (2) were fathers of heresies, and (3) spread prejudices.

Tertullian observed that the philosophers and heretics were constantly raising similar questions. For example: What is the source of evil? Why is evil permitted? What is the source of a human being's existence? And, even, what is the source from which God arose? Tertullian thought that these questions were of no benefit and that the different answers to them spread like cancer.

The philosophers were always searching. But someone sanctified by the Holy Spirit does not need to search any longer, and Christ said to stop searching (*Matthew*, 7:7; *Luke*, 11:9). Tertullian thought that continual searching means that the person has never believed or has stopped believing. He who rejoices in the *Gospel* and believes in Christ does not need to search. What the Christian should believe is already to be found in *Sacred Scripture*, and he who has the fear of God and the true knowledge of God's will has achieved perfect wisdom, even if there are some things he does not know because God has chosen not to reveal them.

Tertullian also engaged in polemics with certain philosophical schools in questions such as our conception of God, creation, and the predestination of the soul. He accused the Stoics and some Greek philosophers (Thales, Heraclitus, Anaximenes, Anaximander, Plato, and Zeno) of treating one or another material element as if it were divine. He criticized Epicurus for making God unfeeling and isolated. He accused the Skeptics of undermining the credibility of our knowledge, which cannot be reconciled with divine providence. Tertullian criticized the Stoics and Epicureans for rejecting the resurrection of the body as impossible.

As he tackled these and other questions, Tertullian willingly or unwillingly entered into philosophical polemics, which he could not have done without some knowledge of philosophy.[4] In the name of faith he was ready to challenge philosophy and human reason in general. So, he writes: "The Son of God died;

we must believe this, because it is absurd. He was buried, but he rose from the dead; this is certain because it is unsuitable, impossible (*credo quia ineptum, credo quia impossibile*)." What he said was later expressed by the phrase *credo quia absurdum* ("I believe because it is absurd"), which accepts the presence of a contradiction to faith at the level of human reason. This would necessarily lead to the theory of two truths later attributed to Averroes (ibn Rushd), that one truth exists for faith, and another, different, truth for human reason.

Tertullian could show the touch of a master when he expressed his aversion to philosophy, and his writing resembles Plato's critique of rhetoric. Tertullian wrote

> What do Athens and Jerusalem have in common? What agreement is there between the Academy and the Church? Christian doctrine comes from the portico of Solomon, who taught that we should seek God in simplicity of heart. Away with attempts to create a Stoic-Platonic-dialectic Christianity! Where is there any similarity between the Christian and the philosopher, between the disciple of Greece and the disciple of Heaven, between a man who aspires to glory and a man who desires life; between a man of idle words and a man of action, between a builder and a destroyer; between a friend and an enemy of error; between someone who corrupts the truth and someone who recovers and teaches the truth?[5]

This flood of rhetorical questions arranged in an Ancient form of parallel expressions in certainly striking. Yet Tatian and Tertullian died outside the Church.

In the Middle Ages St. Peter Damian was one of philosophy's most important opponents. He attacked philosophy in the name of preserving theology. Étienne Gilson has written that Damian was the type of theologian who asked whether the Christian religion contains the whole truth. If it does we do not need anything else. All else is error. The choice between theology and philosophy is the choice between God and the Devil.[6]

Damian thought that secular knowledge does not help to lead souls to God or help us to understand the truth. The work of salvation sows the seed of faith in our hearts and is necessary for understanding the truth. God did not send philosophers and rhetoricians to spread the Gospel. He sent simple fishermen. And a devout life is the road to salvation.

The light of faith is like the Sun. With such light we do not need to light the lantern of learning to see more clearly because faith is sufficient to illuminate everything.[7] According to Damian, the curious desire for knowledge (*cupiditas scientiae*) is the source of all evil and unhappiness; the tree of good and evil in Paradise is a symbol of this desire, and human beings were forbidden to eat of its fruits.

Still, some place exists for secular learning in the life of the Christian. Damian compares philosophy to the golden calf that the Jews adored and Moses

(*Exodus*, 23:20) ground into dust, dispersing the dust into water for the children of Israel to drink. For Christians, they worshipped pagan society as a golden calf, an idol. The river corresponds to the Savior who sows the fire of love in the hearts of the pagans and consumes all forms of idolatry with fire. The calf must be ground into dust because the Devil is the source of pagan societies. By the sprinkling of the dust into the water the pagan will drink, the pagan is won for the Christian society and the Church. The calf was golden because it was built by the sages of this world who pay homage to the Devil.

Damian thought that these sages must have lost their minds if they presented the indestructible God with the help of imaginary images of people or animals. The poets, magicians, astrologers, and all those who studied the liberal arts followed the philosophers. According to Damian that entire body of knowledge had to be ground to dust to incorporate it again to true wisdom. This body of knowledge had to be purified if it was to benefit Christians.[8]

Thereby it was possible to recover pagan knowledge for faith and culture instead of completely rejecting it. But the condition was that theology must occupy first place and would command secular knowledge as a great Lady commands her handmaiden:

Quae tamen artis humanae peritia, si quando tractandis sacris eloquiis adhibetur, non debet jus magisterii sibimet arroganter arripere, sed velut ancilla dominae quodam famulatus obsequio subservire, ne si praecedit oberret, et dum exteriorum verborum sequitur consequentias, intime virtutis lumen et rectum veritatis tramitem perdat.

(Nevertheless, if this expertise in human art is ever applied to sacred eloquence, it should not arrogantly assume for itself a master's right. Instead, just as handmaiden, it should be subservient with some submissiveness of service, lest, if it lead, it might get lost; and, while it follows the consequences of exterior words, lose the most interior light of power and right path of truth.)[9]

In this way Damian wavered between a complete rejection of secular knowledge wherein faith is completely sufficient and a recognition that this knowledge can be useful so long as we completely subordinate it to theology.

Pope Gregory IX officially endorsed these anti-philosophical tendencies in a letter to the theologians of the University of Paris dated 7 July 1228. The Pope wrote that to yield to pagan philosophical teachings and employ them in the interpretation of the *Sacred Scriptures* is an offense and a real aberration. It is the result of vanity, an attempt to glory in learning that shows no concern for the listener.

Professors have the task of lecturing theology in accordance with the tradition of the Fathers of the Church, rejecting everything opposed to the doctrine of God, and subjecting all things to the law of Christ. In this they should rely on God and not seek arms from among the pagans. To make the holy words that come from God agree with the doctrine of philosophers who did not know God is to introduce idols to the temple of the Lord. And if someone wishes to provide arguments for faith with the help of natural reason, he thus makes faith unnecessary, since "faith has no merit if it is proven by reason"("*fides non habet meritum, cui humana ratio praebet experimentum*").

Pope Gregory IX said that we must measure the value of all the sciences by the help they provide to theology.[10] Despite his strong attack on secular knowledge, the Pope, like Damian, ultimately left the door open for this knowledge, although it was as the handmaiden of theology, *ancilla theologiae*.

Thirteen

HANDMAIDEN TO THEOLOGY

To this day when people describe Medieval philosophy as the *ancilla theologiae*— handmaiden to theology—they often intend the term to be derogatory. The term implies that philosophy is not an autonomous science and is of little value, that such philosophy is dictated by theology. When people use this term it is often a disavowal of the intellectual heritage of the Middle Ages. Still, the expression *ancilla theologiae* does not mean what is commonly thought, and Medieval philosophy was truly philosophy and reached a high intellectual level.

As it turns out, this pejorative evaluation of Medieval philosophy was not the work of objective historians of philosophy or something contrived by Marxist ideologues. It first appeared during the Reformation in a campaign against the papacy. An editor wrote in the preface to A. Tribbechovia's *De doctoribus scholasticis* [*About the Scholastic Doctors*] (written in 1665, published in Jena, 1719) that the work could not be published earlier because it disagreed with the Scholastic conception of philosophy as dedicated to the service of papal theology. "[*N*]on puduit asserere scholasticam esse 'philosophiam in servitute theologiae papae redactam.*'"[1]

Tribbechovia thought that the ecclesiastical censors watched to see that philosophical works were in line with official theology. Philosophy as the handmaiden to theology was philosophy subordinated to the theology endorsed by Rome. Different enemies of the Catholic Church, including Marxists, picked up this view, and it eventually became part of common opinion. The propaganda was so effective that, to this day, Medieval philosophy is regarded as the servant of theology and many regard the Middle Ages as the darkest period in history. However, philosophy was treated as the *ancilla theologiae* well before the Middle Ages, and this subordination meant something other than what some people commonly suppose.

In his work *Philosophia christiana cum antiqua et nove comparata* (1878), [*Christian Philosophy Compared with Ancient and Modern*], G. Sanseverino reminds us that Aristotle had already recognized theology as the highest science to which we should subordinate all other domains. Aristotle identified first philosophy and wisdom with theology and said the other sciences were its servants.[2] However, when Aristotle spoke of theology he did not mean a body of knowledge based on supernatural revelation because he did not know of any such thing. He meant metaphysics and an exposition of the first substance, God, as the highest manifestation of being.[3]

The first author who subordinated philosophy to revelation was not a Christian. He was the Jewish, neo-Platonic philosopher, Philo of Alexandria. He applied the term "wisdom" to the *Sacred Scriptures* of the *Old Testament*, to philosophical theology, or metaphysics. And he thought that the proper role of philosophy is to serve revealed knowledge.

Philo's view should not be too surprising, since he was convinced, as was Aristobulus before him, that Greek philosophers such as Pythagoras and Plato had derived their philosophy from the *Sacred Scriptures*.[4] According to Philo and Aristobulus, philosophy originated in the *Sacred Scripturees*. So, all the more, we should subordinate it to *Scripture*.

Philo made an analogy: as the liberal arts (*enkyklia*) are necessary to master philosophy, so philosophy is necessary to acquire wisdom. Philosophy is a training for wisdom. And wisdom is the knowledge and science of the causes of divine and human things. Philo said that "just as encyclic music is the servant of philosophy, so philosophy is the servant of wisdom" (*"hosper he en kyklios mousike philosophias, houto kai philosophia doule sophias"*). Philo illustrated the subordination of philosophy to wisdom (the *Sacred Scriptures*) with the example of the Egyptian slave woman Hagar who was subordinate to Sarah, Abraham's wife (*Genesis*, 16).

Philo influenced Clement of Alexandria. And Clement grafted this conception of philosophy as the servant of theology on to the Christian tradition. Clement drew an analogy between the relation of the liberal arts to philosophy, and the relation of philosophy to wisdom. He concluded that "wisdom is the mistress who rules philosophy" (*"kyria toinun he sophia tes philosophias"*).[5]

St. Gregory Thaumaturgus did not use the term "mistress." He spoke of a helper or companion in labor (*synerithos*), which has approximately the same meaning.[6] Other Greek Fathers, such as Gregory Nazianzenus, Gregory of Nyssa, Amphilochus Iconiensis, and Didymus Caecus, shared this opinion and mentioned the biblical allegory of Sarah and Hagar to give a supernatural dimension to the relation of philosophy to theology.[7]

In the Latin tradition, St. Aurelius Augustine never used the expression *ancilla theologiae*. In the eleventh century St. Peter Damian said that the so-called *artes humanae* should be the *ancilla dominae*. The "human arts" should be "handmaidens of the Lady," and this Lady is the *Sacred Scripture*.

We should remember, however, that when scholars in the eleventh and twelfth century spoke of philosophy, they were thinking of dialectics, not metaphysics, because the strictly philosophical writings of Aristotle had not yet been translated. Dialectics was *de facto* the maidservant of theology. And this, in turn, was in agreement with the Aristotelian understanding of dialectics as an aid (*organon* or instrument). The understanding of philosophy in the eleventh and twelfth century as handmaiden to theology did not mean that *Sacred Scripture* dictated the particular views of philosophy. It meant that theology used the logical instruments of dialectic. Robert Meledunensis wrote: "*non tamen ipsae artes eius (theologiae) sunt ornamentum, sed instrumentum*" ("the liberal arts are theology's instrument, not its ornament").[8]

When metaphysics blossomed in the thirteenth century, its methodological status did not depend on its relation to theology. St. Thomas understood natural theology, or metaphysics, to be an autonomous science whose object is being as being as we know them by the natural light of reason. Metaphysics did not derive its principles from the *Sacred Scriptures* or theology.

This did not mean that theology did not need philosophy. It meant that theology did not impose its principles upon metaphysics. Revelation concerns divine matters. To understand these things, we need philosophy in the sense of metaphysics as the science that rationally explains reality. The authority of the Church is concerned with the criterion of realism as an external requirement whereby it recognizes one order of knowing as capable of serving theology and another as incapable. This does not mean that theology or official authorities can legitimately intrude in the content of philosophical theses.

The expression *ancilla theologiae* is ultimately a metaphor that we cannot analyze simply on the basis of its apparent verbal meaning or biblical context. The proper context is the cognitive and methodological status of philosophy, especially metaphysics, and of natural and revealed theology. Outside this proper context we cannot properly understand the metaphor, and it becomes a tool for ideological manipulation.

Did philosophy in the thirteenth century become an authentic servant of theology as we understand this today, or did philosophy derive its theses from Revelation or papal edict and so fall short of being an autonomous science? To answer the question we must examine the matter first from the point of view of theology and then philosophy.

Fourteen

SACRED SCRIPTURE AND THE PROBLEM OF INTERPRETATION: THE *SENTENCES*

Even the most zealous advocate of the Christian faith who has turned his back on philosophy must eventually face the problem of understanding the *Sacred Scriptures*. While Christians accept the text as sacred, God spoke to us through prophets in human language. And human language is full of words with many shades of meaning. A Christian must interpret the revealed text adequately to render God's thought and message accurately, and so that it does not provide an occasion for heresy. This problem leads us to the question of theology as a science that must provide a correct interpretation of the revelation that Sacred Scripture contains. The concept of science must have the most positive meaning here because theology should provide the proper interpretation, not an accidental interpretation or one based on arbitrary visions.

Scientific theology developed over a long period parallel to the assimilation in the West of the scientific heritage of Greece. During the later Middle Ages, *Sacred Scripture* as the *pagina sacra* (sacred page) started to include a parallel commentary in the form of so-called "sentences."[1] The sentences explained particular words and phrases. They had to draw upon auxiliary fields, such as grammar, rhetoric, history, and law. These domains became collectively known as the *artes liberales* or "liberal arts." This was the only way educationally to advance from reading (*littera*) to understanding (*sententia*).[2]

Use of the liberal arts in the interpretation of *Sacred Scripture* started during the Carolingian Renaissance. In *De schematibus et tropis* Venerable Bede wrote:

Cum autem in sacris paginis schemata, tropi et caetera his similia inserta inveniantur, nulli dubium est quod ea unusquisque legens tanto citius spiritualiter intelligit, quanto prius in litterarum magisterio plenius instructus fuerit.

(Since in the sacred pages we find schemas, tropes and other [rhetorical structures], the reader will grasp the spiritual sense more quickly if a teacher explains the literal sense.)

Similarly, Alcuin of York opposed the use of pagan grammar for the interpretation of the divine grammar.[3]

This critical and analytic tendency, also described as dialectical, lasted until the mid-twelfth century because, during this time thinkers mistakenly understood philosophy in terms of Porphyry's *Isagoge*, Aristotle's *Categories*, Pseudo-Augustine's *Dialectics*, and Cicero's *Topics*.[4]

The distinction of the four senses of the passages in *Sacred Scripture* (literal, tropological, allegorical, and anagogical) was the high point of this approach. After a literal reading the theologian would (1) identify the kind of rhetorical trope associated with a word or expression, (2) interpret the theological meaning of the entire event described, and (3) draw practical conclusions for those who want to travel the road that leads to God. As they followed these four senses, commentaries on *Sacred Scripture* took a hermeneutic character: Peter Lombard's *Sentences* was a parallel commentary that became a classic work.[5]

In the second half of the twelfth century Western scholars became acquainted with Aristotle's *Analytics, Topics,* and *Sophistic Refutations.* These works provided a conception of scientific knowledge that was more than a mere critical analysis of words and concepts.

Soon after translators in Toledo provided Western scholars with the remaining works of Aristotle, including the *Politics, Physics, Ethics,* and *Metaphysics.*[6] This set the stage for theology as a science, a field of knowledge possessing its own proper object, method, and end.

When theologians recognized the proper role of reason in faith, this helped to establish theology as a field distinct from biblical exegesis. Christian thinkers have pondered the proper relation between faith and reason from the time they first reflected on faith. Today, under the influence of Protestantism and the Enlightenment, we are inclined to separate faith from reason and even think of them as opposed. But we should remember that St. Paul said that faith has its own "invisible argument" (*"argumentum non apparentium"*).[7]

St. Aurelius Augustine said that "to believe is to think with assent"— *"credere est cum assensione cogitare."*[8] Faith is not a matter of thought, not sentiment, because, with respect to the truths of faith as truths, our reason cannot know reality directly, as it does in ordinary knowledge. The contents of faith are recognized as the truth held "on faith" that they are so. And this is assent (*assensio*). Augustine adds in one of his letters: "Far from us be the thought that God blames us for that by which he made us higher than other creatures. Far be it from us to suppose that we must believe in such a way as to deny that we must recognize reason or that we need reason; after all, we could not even believe if we did not possess rational souls."[9]

Reason's divorce from faith occurred under the influence of Medieval nominalism, and especially William of Ockham. But, before Ockham, in the early thirteenth century, William of Auxerre wrote of a threefold insight reason brings to faith (*"triplici ratione ostenditur fides"*) with respect to believers, heretics, and those who do not know the faith. Believers can increase and strengthen their faith, heretics can be led out of error, and the ignorant can be inclined to faith.[10] While faith does not cease to be faith, it opens a wide field for reason. Thereby, the work of the reason can extend further than a mere exegesis of the Sacred Scripture.

At the start of his *Summa Theologiae*, Alexander of Hales posed the question: *"Utrum doctrina theologiae sit scientia?"*— "Whether theology's teaching is a science?"[11] He answered that theology is no science. But it is wisdom because, first of all, science bases itself upon relations of cause and effect, while theology concerns the transcendent cause of causes. Second, the aim of science is to learn the truth, while the aim of theology is to evoke in us the love of God. Therefore, theology is more wisdom than it is science. And theology fits in the framework of biblical exegesis.

Alexander maintained that definition, division, and reasoning (*definitio, divisio,* and *collectio*) are the methods of science for discovering the truth. In exegesis we find counsels, examples, admonitions, revelations, and prayers and these lead to love and piety (*pietas*). He adds that human science's method of comprehending truth through human reason differs from divine science's method of using divine tradition to influence us through pitey: "*Dicendum quod alius est modus scientiae, qui est secundum comprehensionem veritatis per humanam rationem; alius est modus scientiae secundum affectum pietatis per divinam traditionem.*"[12]

Alexander did not recognize theology as a science constructed according to the principles set forth in Aristotle's *Analytics*. He presented Peter Lombard's *Sentences* as an aid to biblical exegesis.[13] This was important in the rise of theology as a science because the *Sentences* provided an ordered point of reference when theologians approached some theological problems as theological and not merely exegetical.

Christian thinkers acquired the custom of writing commentaries on the work of Peter Lombard along side biblical commentaries. So, many commentaries on the *Sentences* appeared. Authors arranged these commentaries according to questions, not the chronological order of the books of *Sacred Scripture*. In their structure, the commentaries more closely approached theology as a science.

Finally, St. Bonaventure applied a scientific method in the proper sense, not an exegetical approach, to the problems presented in the *Sentences*. This scientific method entailed investigation and research ("*modus perscrutatorius et inquisitivus*"), while the exegetical method focused on revelation, precepts, prayers, and symbolism ("*modus revelativus, praeceptivus, orativus, symbolicus*").[14]

Bonaventure distinguished among the object of faith (*credibile*) (the First Truth), revealed knowledge (*doctrina sacrae scripturae*), which has the weight of authority, and theology, which considers this revealed knowledge and Truth by reason.[15] Sacred Scripture concerns what we believe by faith as faith (*credibile ut credibile*), while theology concerns the content of faith insofar we grasp it by reason (*credibile ut intelligibile*) and we can subject it to rational investigation (*modus ratiocinativus*).

In the second case, rationality is subordinate to (*subalternatio*), not separated from, faith. The believer, the exegete, and the theologian each have their own methods. But faith must be present in every order, even where we subject faith to reasoning ("*Nisi credideritis, non intelligitis*"—"Unless you believe, you will not understand").[16]

The attempt to construct a theology on such principles as the Aristotelian conception of science demands, presupposes, discovery of analogies with the elements that must be present in the subject of every science. In the case of theology some difficulties existed in discovering such elements.

The structure of the Aristotelian conception of science requires that we start with some first principles that we cannot demonstrate. These first principles are the foundations for demonstrations or syllogisms.

William of Auxerre said that theology possesses principles of this kind: articles of faith.[17] William wrote, "*Sicut aliae scientiae habent sua principia et conclusiones suas, ita etiam theologia; sed principiae theologiae sunt articuli fidei; fidei enim articulus est principium non conclusio.*" ("Just as the other sciences have their own principles and conclusions drawn from them, so does theology; but the principles of theology are the articles of faith, and an article of faith is a principle and not a conclusion.")[18]

We do not draw articles of faith from any prior premises. But an essential difference exists between the principles of the science and theology. The principles of science are evident by virtue of intellectual evidence (*principia per se nota*), while the principles of faith do not possess such evidence for they are principles of faith, not of knowledge as such. At this point, William of Auxerre said that the principles of faith possess their own evidence: the evidence of faith. Furthermore, this evidence of faith is higher than the evidence of knowledge because it comes from God. The evidence of first principles is not diminished because they happen to be principles of faith. On the contrary, they are higher than principles of knowledge.[19]

Thirteenth-century thinkers generally recognized the analogy of first principles in science and theology. In his *Commentary on the Sentences of Peter Lombard* [*Scriptum super libros Sententiarum*], St. Thomas Aquinas said that, from the viewpoint of the act of knowing, we grasp the truth of the principles of science by the light of the agent intellect (*lumen intellectus agentis*), while we grasp the truth of the principles of faith by the light of faith (*lumen fidei*). The light of faith has a role analogous to that of the agent intellect. St. Thomas wrote:

Ista doctrina habet pro principiis primis articulos fidei, qui per lumen infusum per se noti sunt habenti fidem, sicut et principia naturaliter nobis insita per lumen intellectus agentis. Nec est mirum si infidelibus nota non sunt, qui lumen fidei non habent; quia nec etiam principia naturaliter insita nota essent sine lumine intellectus agentis. Et ex istis principiis, non respuens communis principia, procedit ista scientia.

(This science possesses as its first principles the articles of faith, which by an infused light are evident to someone who has faith, just as the principles within us by nature are known by the light of the agent intellect. It is not surprising that the principles of faith are not known to non-believers who do not have the light of faith, for neither would the principles that are within us by nature be known without the light of the agent intellect. This science proceeds from these principles while not rejecting common principles.)[20]

Theology as *sacra doctrina* has elements analogous to those in science: first principles, an intellectual faculty that grasps them, and a light that makes them visible. In the case of theology, these first principles are above our reason. So our intellects alone cannot grasp them. Hence, we need the condition of faith (*habitus fidei*), which is higher than reason. Thomas writes: "[O]nly in the future life when we see God in His Essence will these principles of faith be seen intellectually as self-evident like the first principles of demonstration."[21]

The structure of theological knowledge presupposes subalternation. This means that theology depends upon a higher knowledge, ultimately the knowledge God possesses. "[*S*]*acra doctrina est scientia, quia procedit ex principiis notis lumine superioris scientiae, quae scilicet est scientia Dei et beatorum.*" ("[T]heology is a science that proceeds from principles known by the light of a higher knowledge, which is the knowledge of God and of those who are blessed.")[22]

Subalternation also occurs among the ordinary sciences, where the science of perspective depends upon the principles of geometry, and music depends upon the principles of mathematics. Likewise, the principles of theology depend upon the principles of God's knowledge.

At the same time, faith is clearly internal to theology. For theology is not a purely rational analysis of the meaning of a message. We need first principles to reason in theology. We cannot grasp these principles without faith. So, without faith we could not demonstrate anything. And no science of theology would then exist.

We should keep in mind that the question here is of proof in the Aristotelian sense, where we consider content, not extension. In an extension-based understanding of proof, which is characteristic of nominalism, faith in theology becomes unnecessary. The higher, and subordinate, knowledge must be continuous. The person who possesses a subalternated science perfectly attains science only to the extent that the reasoning in the subalternated science is continuous with that of the higher science: "*Ille qui habet scientiam subalternatam non perfecte attingit ad rationem sciendi, nisi inquantum eius cognitio continuatur quodammodo cum cognitione eius qui habet scientiam subalternantem.*"[23]

Theological knowledge does not aim at drawing conclusions from premises. Ultimately it aims at knowing God to the extent that He has revealed Himself to us.[24] One reason why theology is *sacra doctrina* is that a constant link exists in it between human knowledge and revelation. If theology were separated from revelation and faith, it would no longer be *sacra doctrina*. In theology, the rational development of the data of revelation is distinct from faith. But the work of reason does not take away the supernatural character of truths of faith. Instead, reason is immersed in this supernatural character by faith. St. Thomas wrote, "*Unde illi utuntur philosophicis documentis in sacra scriptura redigendo in obsequium fidei, non miscent aquam vino, sed convertunt aquam in vinum.*" ("Hence, those who use philosophy in the interpretation of scripture in the obedience of faith do not mix water with wine, but they convert water to wine.")[25]

This point leads us to the question of how philosophy and theology connect. If we need rationally to investigate *sacra doctrina*, we must draw upon the field of knowledge best suited for this purpose. This field is philosophy, and especially the philosophy's most important part—metaphysics.

Fifteen

METAPHYSICS AND NATURAL THEOLOGY

Aristotle was not the author of the term "metaphysics." He spoke of "first philosophy," "wisdom," and "theology." Philosophical controversy has existed for a long time about the unity of metaphysical knowledge and the relation of metaphysics to theology in connection with these terms.[1]

We find the key text in Aristotle where theology and first philosophy appear to be interchangeable, in Book 6 of the *Metaphysics* where Aristotle lists the theoretical sciences and mentions physics and mathematics. The last science he mentions is the "first science," "theology," and "first philosophy." He bases his on a difference in the objects of the sciences.

Physics concerns substances that possess in themselves a principle of motion and rest. Physics studies substance under the aspect of form in connection with matter.

Mathematics studies substance with no regard to motion and in separation from matter. Aristotle leaves open the question whether such a substance actually exists or whether this is a question of how we apprehend material substances in mathematics (by what would later be called "mathematical abstraction").

The problem of the object of mathematics relates to the fact that Plato presented a middle sphere, of numbers, between the sensible world and the World of Ideas. Aristotle shows that such a sphere does not exist and that the object of mathematics is the result of a special kind of abstraction.[2]

The first science, theology, or first philosophy concerns unchanging substances, substances that are not in motion and can exist separate from matter, immaterial substances. Aristotle called this science theology, because what is immaterial and unchanging is divine. He identified this theology with first philosophy because the divine substance is the first substance and the first being. So, theology concerns the first being in the hierarchy of being.

This first being and first substance functions as the first cause for all being. Consequently, the study of the first substance is simultaneously the study of first causes. Since theology concerns the first being, it studies a universal body of knowledge and concerns being as being.[2]

This condensed passage from Book 6 of the *Metaphysics* presented many difficulties for commentators. Theology appears as one of the theoretical sciences, but its object is separated substance. Since this substance is the first substance and the cause of all other substances, theology has a privileged position among the sciences. It is not merely one of the sciences.

Also, because theology is identical with the science of being as being, in some way, theology is universal in character.

Furthermore, we do not know the divine substance directly. We know it indirectly as a cause because, in the order of human knowledge, we learn of the first substance through knowledge of material substances.

Thus, as the science, we must link theology concerned with the first substance to first philosophy as the science of being as being. Aristotle's analysis of substance in Book 12 starts by considering the motion and structure of material substances. It does not start from a direct knowledge of the divine substance. This consideration involves the divine substance's existence and nature. Aristotle shows divine substance is to be eternal, in act, necessary, unextended, immaterial, simple, and self-thinking. Its existence is necessary since we must find an adequate cause for eternal motion.

Aristotle's natural theology is not based upon a revelation from God or a direct knowledge of God. It is based upon a metaphysical analysis of the sensible world. Thus, theology is not an autonomous scientific or philosophical discipline. But it does occupy a special position in metaphysical knowledge, because what is divine is also the highest substance.

Aristotle sees the object of natural theology primarily as a cause, being as it is known by the senses, not as a substance. Since scientific knowledge is knowledge by causes, theology is a type of scientific knowledge distinct from mythology.

Aristotle's demythologization of theological knowledge is one of the most important accomplishments of Greek thought. Before Aristotle, Xenophanes had already made a rational critique of anthropomorphic conceptions of the gods. But he did not develop his critique into a definite methodology of scientific knowledge. Aristotle started the scientific treatment of divine being.

The conception of God or the Absolute in natural theology depends upon the conception of being developed in metaphysics. So natural theology has its basis in metaphysics.

During the Protestant Reformation and later in the Enlightenment, some thinkers criticized the approach of subjecting revelation or Sacred Scripture to Greek philosophy and reason. The critics thought that the Aristotelian approach would lead to a distortion of Christianity.

However, from the point of view of the culture of science, this approach caused something amazing to take place. For the first time in history human beings subjected to the definite and rigorous intellectual criteria demanded by scientific knowledge what people had previously considered in mythology and poetic fantasies. The Ancient Greeks took the first step when they searched for God in *theoria*.

Medieval thinkers also took an important step when they marked out the framework for interpreting the revealed message. They considered our human way of knowing the world and the structure of the reality we know. Without losing its supernatural dimension, in this way, theology developed with the highest domain of human culture: science in the broad sense of knowledge of the truth.

In one way or another, directly or indirectly, interpretation of *Sacred Scripture* must appeal to some conception of human knowledge and science. The *Sacred Scriptures* speak of revealed things that we cannot know in the same way as God and the angels know them. We must properly prepare, or train, human reason to receive, understand, and acknowledge these truths as well as possible. Theology's task is to receive revealed truths properly. This requires that the human knower apply refined skills in knowledge: science in general and philosophy in particular.

Philosophical reflection also enables us to define the limits of natural human knowledge in matters mentioned in *Sacred Scripture*. Paradoxically, human reason can establish its own measure to determine that revealed truths are revealed truths, human fiction or something more properly treated as an object of natural science. The great accomplishment of the scientific culture of the Middle Ages was to discover the proper place in matters of faith for human reason as perfected by science.

Sixteen

PHILOSOPHY IN THE SERVICE OF THEOLOGY

Philosophy is not related to theology in such a way that theology dictates what philosophy should teach. Theology needs philosophy in the interpretation of divinely revealed truths. The arts of the *trivium* are helpful tools in philosophy. But this does not mean that philosophy dictates the rules of grammar and logic. Likewise theology uses philosophy. But it does not influence what philosophy teaches.

We can say what type of philosophical method and teaching is useful in theology and what type is useless or destructive. If some popes have made authoritative statement on philosophy, they were concerned with the relation of philosophy to theology and spoke out of concern for the faith. But they did not violate the autonomy of philosophical knowledge.

St. Paul provided a basis for the rational analysis of the content of faith when he spoke of faith having its own "invisible argument" ("*argumentum non apparentium*," *Hebrews*, 11:1). Marie-Dominique Chenu writes that faith is more than a conviction. It has its own rational argument that we must extract. The rational element in faith allows our reason to penetrate faith and opens the field for philosophy. Philosophy as an established science assists theology. The reason for bringing philosophy into theology is the content of faith insofar as we can understand it (*credibile ut intelligibile*), since a mere exegesis of texts at the level of the "liberal arts" is insufficient.

All the more, no reason exists for theology to reject philosophy's theses. We know this because philosophy reveals the natural privation that requires grace, and philosophy proves that no such oppositions exist between faith and reason and nature and grace.

No contradiction exists between reason and faith or nature and grace. The supernatural brings the natural to completion.[1] St. Thomas Aquinas explained that, while theology relies upon faith's light, philosophy relies upon reason's natural light. And the two cannot be contrary to each other. He stated:

Sicut autem sacra doctrina fundatur super lumen fidei, ita philosophia super lumen naturale rationis; unde impossibile est quod ea quae sunt philosophiae sunt contraria eis qui sunt fidei, sed deficiunt ab eis.

(Just as theology is founded upon faith's light, so philosophy is founded upon the light of natural reason. So, while what is of philosophy falls short of what is of theology, it is impossible that those things that are of philosophy are contrary to those that are of faith.)[2]

As St. Thomas said, theology can use philosophy in three ways. (1) Philosophy demonstrates some things that are an introduction to theology (preambles to faith) and the things that we can learn about God by natural reason (for example, that God exists and is one). (2) By likenesses drawn from the world, theology can use philosophy to make matters of faith better known. (3) Theology can use philosophy to refute views contrary to the faith.

> *In sacra doctrina philosophia possumus tripliciter uti: Primo ad demonstrandum ea quae sunt preambula fidei, quae necessaria sunt in fidei scientia, ut ea quae naturalibus rationibus de Deo probantur ut Deum esse, Deum esse unum, et huiusmodi, vel de Deo vel de creaturis, in philosophia probata quae fides supponit. Secundo ad notificandum per aliquas similitudines ea quae sunt fidei, sicut Augustinus in libro de Trinitate utitur multis similitudinibus ex doctrinis philosophicis sumptis ad manifestandum Trinitatem. Tertio ad resistendum his quae contra fidem dicuntur, sive ostendendo esse falsa, sive ostendendo non esse necessaria.*[3]

When theology uses philosophy, theology can draw conclusions from the articles of faith, and, thereby, the believer can know faith's implications. As St. Thomas says, *"Fidelis potest dici habere scientiam de his quae concluduntur ex articulis fidei."* ("We can say that the faithful have knowledge of what they conclude from an article of faith.")[4]

Mieczysław Albert Krąpiec has studied this problem from the metaphysical viewpoint before the publication of the encyclical *Fides et ratio*. Krąpiec wrote that the theologian who explains and interprets revelation and the research scientist who, as a scientist, utilizes accepted scientific presuppositions or develops new ones in their light work differently. The research scientist has no need completely to verify the truth of scientific presuppositions. The theologian, however, cannot use scientific hypothesis that might later turn out to be wrong because the theologians work for the sake of human beings and our most sublime "'divine' experiences." The theologian constantly deals with the life of real human beings "who in their daily conduct must be found in a normal and real state in relation to God, not in some imaginary world created by writers of fantasies, haunted people, and oracles, and so on."

Real human beings live in, and benefit from, the real, not an abstract, world. Thus, in a minimal sense, a theologian must accept this world and consider it when studying the relation of human beings to a God who reveals teachings through his prophets, apostles, and Jesus Christ.

Consequently, Krąpiec concluded, "philosophical systems that put their own abstract meanings in place of the world as the object of our understanding of the essential context human life cannot serve as a foundation for theological studies."[5]

To sum up, the conditions that entitle the theologian to analyze the content of Revelation, we should see that these analyses cannot be performed in isolation from the essential elements and conditions of knowledge that are the reality of the really existing world, the reality of the human being as the one addressed by revelation, and the character of human language as the semiotic vehicle of the contents of knowledge.

In 1981, in the encyclical *Fides et ratio*, Pope John Paul II confirmed the position of Ancient and Medieval authors such as Sts. Augustine, Bonaventure, and Thomas Aquinas regarding the relation between philosophy and theology. The Pope recalled their arguments as being continually relevant and true. The title of one chapter is especially striking: "The Enduring Originality of the Thought of Saint Thomas Aquinas." The Holy Father repeatedly spoke of the need to return to realistic classical philosophy as the philosophy that, in a sound way, can serve theology. Hence, he wrote:

Profoundly convinced that "whatever its source, truth is of the Holy Spirit" ("*omne verum a quocumque dicatur a Spiritu Sancto est*") Saint Thomas was impartial in his love of truth. He sought truth wherever it might be found and gave consummate demonstration of its universality. In him, the Church's Magisterium has seen and recognized the passion for truth; and, precisely because it stays consistently within the horizon of universal, objective and transcendent truth, his thought scales "heights unthinkable to human intelligence." Rightly, then, he may be called an "apostle of the truth." Looking unreservedly to truth, the realism of Thomas could recognize the objectivity of truth and produce not merely a philosophy of "what seems to be" but a philosophy of "what is."[6]

Part Four

SCIENCE: TOWARD
TECHNOLOGY AND IDEOLOGY

St. Albert the Great (1200–1280), Catholic University of Lublin

Seventeen

THE PROBLEM OF THE CONTINUITY OF SCIENCE IN THE MIDDLE AGES

Herbert Butterfield has said: "[W]hen we speak of Western civilization being carried to an Eastern country like Japan in recent generations, we do not mean Graeco-Roman philosophy and humanist ideals, we do not mean the Christianizing of Japan. We mean the science, the modes of thought and all the apparatus of civilization which were beginning to change the face of the West in the latter half of the seventeenth century."

The Renaissance brought with it a new conception of science that, with some modifications, is still with us today. Its motto is utility. But this utility has a broader sense than in ordinary speech. Science should give us mastery over nature, the world, even the cosmos. It should improve and perfect human beings and society.

This great task could not have been the result of a merely internal evolution in the conception of science. The Renaissance re-Easternized science from knowledge for knowledge itself (*scire propter ipsum scire*) to knowledge for use (*scire propter uti*) and involved the context of civilization as a whole beyond the boundaries of science.

Some authors treat science's evolution on a microscopic scale or as something that happens completely within the confines of science. In this approach they ignore important reasons for science's changing place within culture.

Change in purpose entails a change in method, in the formal object, and in the material object of scientific research. Such change is crucial. This change in purpose arose from science's external context and found fertile ground there. Thinkers outside science became aware that they could transform our understanding of science. This change was of such importance and influence that no marginal cause could have dictated it. Science changed when civilizations came into close contact, and especially as other civilizations pressed upon Western or Latin civilization.

Utilitarianism in the material sphere led to the transformation of science into technology. In the human sphere it led to the transformation of philosophy into ideology. Technologization of the pure sciences and ideologization of the humanities are effects that we still feel today. And they are still advancing.

The new understanding of science has become the most characteristic feature of the West. Science and technology are now considered to be the West's contribution to world culture, while morality, art, religion, the Greco-Roman heritage, and Christianity count for little. The West can easily influence other lands today in the areas of technology and ideology, but not in culture in the broad sense.

What was the context in which the conception of science in Western culture changed from *scire propter ipsum scire* to *scire propter uti*? Answers to these questions are crucial in showing the West's identity and in defining the paradigm of science as an essential element in this identity.

To some degree, perhaps, this change was the result of an internal evolution in the conception of science. But it is certainly linked to the propagation of specific Eastern ideas that later took on a Western form.[1]

The Romans had no problem assimilating the Greek heritage because many Romans were equally fluent in Latin and Greek. Their Greek teachers introduced the Romans to a higher culture. Since many Romans knew Greek, few Greek works were translated into Latin. When the Empire was divided into the Eastern and Western knowledge of the Greek language and culture started to decline in Rome. An urgent need then existed to translate Greek works into Latin. The decline and fall of the Western Empire was a period of intense translation on a grand scale. This made possible the Romanization of the Greek heritage.

Chalcidius lived in the fourth-century A.D. and translated Plato's *Timaeus*. His translation was also popular during the Middle Ages. Anicius Manlius Severinus Boethius (480–524) translated some of Aristotle's logical works (the *logica vetus*), Euclid's *Elements*, and Porphyry's *Introduction to Aristotle's Logic*. He also wrote several textbooks on the liberal arts based upon Greek sources.[2] Up to the twelfth century, the only works of Aristotle known in the West were the *Categories* and *On Interpretation*.[3]

Cassiodorus (480–575) also contributed to the preservation of Western culture. In his cloister he organized translations. He regarded knowledge of the seven liberal arts (grammar, rhetoric, logic or dialectics, arithmetic, geometry, astronomy, and music) as an essential part of his monks' education. He added to these his own manual based on primary Greek texts.[4]

Isidore of Seville (560–636) wrote two especially influential works: *On the Nature of Things*, and *Etymologies*, which were an encyclopedic compilation of the whole body of Greek and Roman learning.[5] Venerable Bede (673–735), Alcuin of York (735–804), and Rhabanus Maurus (776–856) were other authors who helped preserve the classical heritage).

After the Roman Empire's fall the classical tradition only the monasteries and monastery schools preserved classical culture. St. Benedict founded the first such monastery in Monte Cassino in 529 A.D., the same year in which the emperor Justiniian dissolved Plato's Academy, which had been in existence for almost a thousand years. The professors of the closed Academy found shelter in the court of the Persian king and after a few years returned to Athens. The Platonic Academy was never re-opened, and the Academy in Florence that was founded about 1,000 years later was an artificial continuation of the original. The year 529 marks the end of the Ancient world and the start of the Middle Ages.

While the West struggled for centuries to preserve its civilization against the attacks of the Goths, Vandals, Franks, and later against the Normans, the Eastern Empire grew in strength. The Byzantines had a low regard for the Latins and their language.

They spoke Greek and had direct access to the heritage of the Ancient world. From the fourth century to its fall, Byzantium primarily spoke Greek. The Byzantines deliberately eliminated expressions borrowed from Latin.[6] Christianity was the State religion. Byzantium treated pagan literature as something alien. The Byzantines were very cautious in their use of it.

Byzantine education had three degrees: (1) basic learning (*enkykles paideia*); (2) literature, history, geometry and geography, grammar; and (3) higher education (rhetoric and philosophy).[7] Alexandria, Caesarea, Gaza, Antioch, Ephesus, Nicea, and Edessa were some large centers of education that belonged at different times to the Byzantine Empire. Constantine probably founded a secular palatine school in Constantinople. It lasted until the fall of the empire in 1453. There were also cloister schools and the Patriarchal school. Many of these centers were destroyed by the invasion of the Arabs. In the twelfth century, Western scholars started to learn from the schools that remained, such as the palatine school in Constantinople.

The Byzantines did not develop the natural and mathematical sciences, and they reduced philosophy to commentaries on classical authors. Themistius (born around 385), Simplicius (died after 533), and John Philoponus were well-known commentators.[8]

The conquests of Alexander the Great had spread Greek science and learning widely throughout Asia to the Indus river. The most active centers of learning were Alexandria in Egypt and Bactria in Central Asia.[9]

The Syrians played a special role in transmitting Greek culture. The East considered the Syrian language an international language. Persians, Byzantines, and Arab used it.[10] The Syrians translated important Greek works into Syrian and so made them available to other nations in the East. They translated Aristotle's *Organon*, *Poetics*, and *Rhetoric*, works of Plutarch, Lucian, and pseudo-Socratic dialogues.[11] Plotinus' famous disciple Porphyry was of Syrian descent. His *Isagoge* was translated many times from Greek to Syrian.[12]

A tradition of translations from Syrian and Greek into Arabic started in the mid-eighth century in the caliphates during the reign of al-Mansar. His grandson Harun al-Rashid brought manuscripts from Byzantium. Al-Rashid's son, al-M'mun, founded the House of Wisdom in Baghdad, and this became the chief centre for translation. Hunayn ibn Ishaq (808–873) directed of the House of Wisdom. He was an Arab and a Nestorian. He was equally fluent in Greek and Syrian. He brought together translators in Greek, Syrian, and Arabic who compared different manuscripts and studied the meanings of the translation.

They translated medical words (Hippocrates and Galen), philosophical works (three dialogues of Plato [including the *Timaeus*], writings of Aristotle [including the *Metaphysics*, *On the Soul*, *On Generation and Corruption*, and portions of the *Physics*]), different logical and mathematical works, and a Syrian version of the *Old Testament*. The Arabs were quick to make use of the translations of almost all the works on Greek medicine, natural philosophy, and mathematics.[13]

New Arab centers of learning appeared up to the thirteenth century. The leading schools were in Baghdad, Cairo, Cordoba, and Toledo. The Christians captured Toledo in 1085, Cordoba in 1236, and Seville in 1248. Europe, thereby, received much of the classical culture and Arab accomplishments. The invasion of the Mongols and their occupation of Baghdad in 1258 marked the start of the end of this high Arab culture.[14]

Several works in mathematics and on the astrolabe were translated from Arabic to Latin in the tenth century. In the eleventh century a monk called Constantine, from North Africa, entered Monte Cassino and translated from Arabic medical works, including Hippocrates and Galen. In the twelfth century Spain was the center for translations to meet the needs of the Latin West. Some scholarly centers and their libraries fell into the hands of the Christians. Many translators (such as John of Seville, Hugo of Santalla, and Mark of Toledo) were Spaniards. Others (such as Robert of Chester, Herman of Dalmatia, and Plato of Tivoli) came from other lands. Gerard of Cremona (1114–1187) was the greatest translator. He translated twelve works on astronomy, seventeen on mathematics and optics, fourteen on logic and the philosophy of nature, and twenty-four works on medicine: seventy to eighty works in all.

Boethius translated some works from Greek into Latin in the sixth century, as did John Scotus Eriugena in the ninth century. But translation on a large scale only started in the twelfth century. The chief center for translation then was in southern Italy. Ancient Greek colonies existed there and maintained contact with Byzantium. Jacob of Venice translated several works of Aristotle and works on mathematics and optics. In the thirteenth century William of Moerbeke (1260–1286) translated the complete works of Aristotle, neo-Platonic commentaries on Aristotle, and the mathematical works of Archimedes.[15]

From this cursory review of the paths of Greek and Roman scientific culture, we can see that this culture did not proceed in an unbroken line. The eleventh, twelfth, and thirteenth centuries were a watershed where elements of different civilizations met, including Egypt, Persia, Syria, Byzantium, and the Arab world. After this watershed, return to the path of Greek *theoria* would not be easy and more than one current of thought would come to exist in the reborn Western culture in Europe.

The fall of the Western Empire influenced the break-up of Western culture as a whole, and science was part of that culture. In the Eastern, or Byzantine, Empire the continuity of culture was preserved to a much greater degree than in the West. And the Greek and Hellenic scientific heritage became part of the new civilization that arose in the seventh century with Mohammed's new religion. The Arabs were intermediaries in the rebirth of Western culture in the eleventh and twelfth centuries. The Latins received the scientific treasure of the Ancient world from the Arabs. In 1204, the Crusaders conquered Byzantium, and, in the thirteenth century, many works were brought west from Byzantium.[16]

The West did not recover a link to its own culture in one fell swoop. It recovered it in stages. And the recovered heritage included borrowings from the cultures that transmitted this heritage.

Later thinkers often regarded these borrowings and accretions as part of the original Greek culture. They ascribed works from other civilizations to authoritative Greek authors, such as Plato and Aristotle. While such works were more widely received, this practice also led to important changes in the conception of science. While Scholastic philosophy as represented by St. Thomas Aquinas was continuous with the works of Aristotle, the ideas of Roger Bacon, considered to be the precursor of the modern conception of science, had many elements that were not Greek or Latin. Roger Bacon was not completely aware of this because he had been greatly influenced by a work that, at the time, was falsely attributed to Aristotle, the pseudo-Aristotelian *Secretum secretorum* [*Secret of Secrets*].

After centuries, the process of recovering the heritage of the Ancient world was completed in the late Renaissance. The works of Archimedes and the mathematicians of Alexandria became available in translation in 1543. Butterfield maintains that, during the Renaissance, Archimedes' works, translated in 1543, "represent the last pocket of the science of antiquity which was recovered in time to be an ingredient or a factor in the formation of our modern science."[17] Another century would have to pass for the new conception of science fully to crystallize.

Eighteen

THE PROBLEM OF EXPERIMENTAL SCIENCE IN THE MIDDLE AGES

Today, common opinion maintains that leading characteristics of modern science are the role of experiments, inductive method, and mathematical treatment of scientific knowledge. For a long time people thought that this conception of experimental, inductive, and mathematical science arose in the sixteenth and seventeenth century. And, so, this science belonged to modern times as opposed to the dark Middle Ages.

While the understanding of science's history changed in the late nineteenth century, this opinion lingers today. Largely because of a work by William Whewell (*History of the Inductive Sciences from the Earliest Times to the Present Time*), historians of science recognize Roger Bacon (1214–1294) as the precursor of the modern conception of science. Whewell's work presents Bacon as someone unique in his time because of the importance he attached to experiment and the inductive method.[1]

Over time historians of science such as Pierre Duhem, L. Thorndike, and R. Carton have shown that Bacon was not a Medieval exception because philosophers such as Robert Grosseteste, William of Auvergne, Peter of Spain, and Albert the Great also emphasized the role of experiments in physical science. Historians have also shown how, in the Hermetic tradition, of which Bacon was a diligent student, alchemy and magic played an important role in the formation of a new conception of science.[2] Hermetic writings, especially a mysterious work called *Secretum secretorum* [*Secret of Secrets*] strongly influenced Bacon. This work was held in high regard because it was attributed to Aristotle. Bacon added a commentary to it in the form of a gloss.

Aristotle allegedly addressed the *Secretum secretorum* to Alexander the Great. It says that a body of knowledge exists that mainly secrets and mysteries can transmit. In Ancient times God revealed this knowledge to prophets and those he chose, and enlightened their minds. Adam, his son Seth, and Enoch possessed this knowledge. And some writers identified Enoch with Hermes.

Ancient Greek mythology celebrated Hermes, also called Hermes Trismegistus, for mystic and divine knowledge. The Hermetic tradition taught that Hermes and Aristotle never died and that Aristotle recognized the doctrine of the Holy Trinity. The hermetic views in the *Secretum Secretorum* were a syncretic blend of Greek philosophy and Christian religion.[3] In this way, Aristotle and Christianity gave credibility to Hermeticism and made Hermeticism easier to become part of Western thought and internally modify the typical Western concepts associated with scientific knowledge.

1. Utility

In his well-known work *Mediaeval Science and the Beginnings of Modern Science*, Alistair Cameron Crombie states that Arabic thinkers "made their most important and original contributions to the history of European science in the fields of alchemy, magic, and astrology." Crombie maintains that the reason for these contributions is partly the result of a special way of approaching problems in physics "characteristic of the living tradition of Arab thought."
     According to Crombie, most important in this tradition was for a person to come to know nature to dominate it. In this respect, Crombie maintains, traditional Arabic thought differs from the European thinker's inclination "to seek in nature the facts that best illustrate the moral designs given by God or to seek rational explanations for the facts described in the *Bible* or those seen every day in the world." Instead, Arabic "researchers desired to find the elixir of life, the philosophical stone, a talisman, a word of power and the magical properties of plants and minerals."[4]
     As a result, the role of the Arabs as mediators in the assimilation of Greek science by Medieval Europe caused some practices alien to science, such as magic and alchemy, to creep into science. These elements had been continually present in Eastern culture, which had no knowledge of science as science. Moreover, love for truth or knowledge as such did not dictate the presence of occult knowledge in Eastern culture. Supposed efficacy for gaining mastery over nature and the world caused its presence in Asia Minor and the Middle East. In this way a new end of scientific investigation, utility, became mixed with magic and alchemy.
     Bacon was a diligent student of Arab thought. In a letter to the Pope in 1267, he wrote that, while he had studied the sciences since he was young, the last twenty years were especially important. Over those years he had spent over two thousand pounds on occult books, experiments, tables, and so on.[5] This period also had an important influence on Bacon's views about scientific knowledge. He started to appreciate science's practical ends, and he made these the primary ends.
     Bacon distinguished different kinds of alchemy according to different degrees of utility. He noted the existence of "another alchemy that is effective and practical and teaches how to produce artificially noble metals, pigments and many other things better in greater quantity than does nature." And he stressed, most importantly, "that this kind of knowledge is more significant than all that was known before because it provides great benefits."
     Bacon continued his lofty praise of this useful alchemy: "Not only can it provide wealth and many other things necessary for the general good, but it also teaches skills such as how to prolong human life beyond the term foreseen by nature."[6] This fragment certainly contains the credo that, during the sixteenth and seventeenth centuries, would become the foundation of the modern conception of science: the most significant knowledge is that which provides the most

benefits, and that which provides the most benefits is thereby the most useful.

2. Experiment

The term "experiment" comes from the Latin noun *experimentum*, derived from the verb *experior*, *experiri*, which means, "to test" and "to experience." The corresponding Greek term is *empeiria*, and derivative terms, in all European languages, as in English when we speak of "empirical data." We can translate *empeiria* as "experience," "skill," and "acquaintance." The distinction we make today between experience in the receptive sense and experimentation as an action involving careful planning first appeared during the Renaissance.

Plato and Aristotle wrote of *empeiria* in the contexts of art as a skill for producing things, the origin of general ideas, and acquisition of knowledge. In the dialogue *Gorgias* Socrates said: "O Chaerephon, there are many arts among mankind which are experimental, and have their origin in experience, for experience makes the days of men to proceed according to art, and inexperience according to chance."[7] Thus Plato held that art arises by experience. In the *Timaeus* experience cannot provide a way for human beings to perform the actions of God; God alone can join and divide the mixtures that underlie the construction of the cosmos.[8]

In the *Posterior Analytics* Aristotle put experience before art and organized knowledge. Experience has an aspect of generality that distinguishes it from singular impressions and from memory in which we repeatedly grasp the same object. He said: "And again from experience, that is, from generalities retained in their wholeness in the soul, unity alongside plurality, which is simultaneously one and the same in its plurality, comes that which is the principle of art and organized knowledge; art is directed to that which comes into existence, and organized knowledge is directed toward what exists."[9]

The Polish translator of Aristotle's work, Kazimierz Leśniak, uses the term "exists" to translate the Greek verb *einai*. I would use the term "is" to translate *einai* because Aristotle was not writing about the existential meaning, which he did not see. He was writing about the essential meaning. If Aristotle says that something is, he is thinking of it as permanent and invariable.

Aristotle expressed the same view, somewhat modified, in his *Metaphysics* where he associates experience more strongly with concrete things. While experience underlies art and organized knowledge, we find generality only in art and knowledge, not in experience:

Experience arises in men from memory: the repeated remembering of the same thing finally is transformed into one experience. Experience even seems to be like science and art. But they differ in that people acquire science and art by experience. For as Polos rightly says experience produced art and inexperience produced accident. Art arises when from many empirical apprehensions one generality is produced which refers to instances similar to each other. It is merely experience when someone says that a certain thing helped Callias, Socrates, or someone else in a certain illness in particular cases. To say that in a particular illness a particular remedy helps all those who are described by one concept such as phlegmatics, cholerics, or digestive upset, is art. In the case of act on experience seems not to differ in any respect from art. Moreover he who has experience performs something better than he who has understanding without experience. The cause lies in the fact that experience is knowing particular cases while art is knowing that which is general.[10]

Aristotle showed how experience connects with, and differs from, art and organized knowledge. He linked experience with concrete cases and art and organized knowledge with generality and necessity in judgment. We do not know generality and necessity *a priori*. We know them with the help of experience. And experience arises from sense impressions and memory.

Experience allows us to form a general perception and unified image about particular cases. Forming such a general image is a necessary element in the rise of scientific knowledge because scientific knowledge involves making universal and necessary judgments based upon general perception. Experience here is not yet experiment as we mean it today. We may wonder how and why the concept of experience has changed. To find the answer we must take a broader look at the matter.

The lands and peoples in the empire of Alexander the Great became Hellenized but did not thereby lose their own cultures and former practices. A syncretic mixing of civilizations occurred. The East was Hellenized and Hellenism was Easternized. The cultures of Babylon, Mesopotamia, Syria, Palestine, Egypt, and Persia predated Greek culture. They were subject first to the influence of Hellenism, then to the Romans. And, during the seventh century, they came under Arab rule.

In the Eastern world alchemy stood in the place of theoretical science. It developed especially in the Arab world between the ninth and eleventh centuries. Alchemists performed different experiments to get different grades of metal, and their ultimate goal was to produce gold. They also wanted to make pigments and to find the philosopher's stone.

An alchemical tradition developed and the alchemists published descriptions of their experiments.[11] In the Christian West, starting in the ninth century the mechanical arts developed. The Europeans were interested in new inventions to save labor in agriculture and in workshops. And they found ways to bring increased mechanization to mining and the extraction of minerals.[12] While they performed experiments they did not yet take a scientific approach in experimentation.

Abedelhamid I. Sabra notes that the concept of the experiment as distinct from the concept of experience appears in the work of Ibn al-Haythama. "*I'tibar*" was translated from Arabic to Latin as *experimentatio*. Some precedent existed for this translation. In the *Almagest* and the *Optics* Ptolemy had written of a test of confirmation in astronomy. Ibn al-Haythama wrote of confirmation in other sciences, especially in optics.[13] We can find a broader understanding of the term "experience" in the writings of Roger Bacon where it includes the Aristotelian meaning and the meaning it had in the works of Ibn al-Haythama.[14]

The concept of experience as aimed at the verification of a theory developed gradually. Crombie writes that the Ancient Greek mathematicians derived their first perceptions from which they "discovered the abstract order that stands behind the chaos of immediate experience." The first perceptions the Greek mathematicians discovered from the sphere of simple relations that hold true in mathematical sciences like astronomy, optics, mechanics, and acoustic the abstract order that stands behind the chaos of immediate experience.

Crombie maintains that the Greek mathematicians attempted to explain a phenomenon by seeking the smallest number of simplest principles that would be its cause. Presupposing these principles, these mathematicians recognized that a particular phenomenon must occur. In this way they used geometry's speculative power and imposed upon phenomena geometry's "deductive logical structure and its model that defined a particular shape in space for every phenomenon."

According to Crombie:

Euclid and the other Greek mathematicians wanted to extend their field of investigation to different phenomena in a purely theoretical way within their geometric or arithmetic model. They later became aware—as Ptolemy did— that in the investigation of complex phenomena, the postulates must be controlled by observations and experiments to determine whether a possible theoretical model leads to results corresponding to the real world. In this way the style of scientific proof in optics began to be seen in the fourteenth-century West—chiefly because of Ptolemy and later because of Alhazen— as a style of empirically controlled presuppositons.[15]

The concept of experiment characteristic of the modern conception of science is not completely original. It has its origins in Ancient mathematics and its model for interpreting the world. The purpose of an experiment is to confirm a model built on some set premises.

Still, we should not forget that in the thirteenth and fourteenth centuries, the concept of experiment was interchangeable with that of the secret (*secretum*) and secrets were recipes and rules shrouded in mystery. Many popular books of secrets were published in the fifteenth and sixteenth centuries, though they circulated outside the official university circles. Experiments, or secrets, were not part of the scientific canon of the time. But people found them attractive because they promised results and opened new possibilities.[16]

From this conception of the experiment or secret rose the concept of experimentation as an essential element of a new conception of science.

Experimentation was a road along with mathematical models that led to inventions.

3. Inventions

The mathematical conception of the experiment opened the field to new models of knowledge and inventions. The Greek cult of science for science's sake had the result that the Greeks' intellectual potential bore fruit more in the pure sciences than in many great inventions.

Still, the Greeks had excellent architects who built many beautiful temples, amphitheatres, even entire cities. And they had many inventive builders and inventors. Plato's friend Archytas of Tarentum built a rattle for children, a mechanical bird that imitated the motion of a real bird and was driven by compressed air, and probably invented the screw.[17] In the fifth-century B.C., the architect Phaiax built a system of canals and reservoirs to supply Acragas with fresh water.

Herodotus wrote of the three greatest engineering achievements of the Greeks on the island of Samos. (1) The Megarian Eupalinos was the major builder of a mountain tunnel 1,260 meters long, nine meters high and 90 centimeters wide. The tunnel brought fresh water to the city. (2) The Greeks constructed a high dam was constructed on the sea around the port. (3) Rokos built the largest temple in the known world.

The Spartans invented a new type of lock and key.[18] The tyrant Dionysius the Elder, who ruled Syracuse in 399 B.C., gathered specialists to provide inventions especially for military use. They developed a catapult and other new weapons, projectiles, and large ships (*tetreras*). During the early fourth-century B.C., a Pythagorean philosopher, Zopyros of Tarentum, invented the catapult and the crossbow. Straton, a disciple of Aristotle, wrote a work called *Mechanika* that was erroneously attributed to Aristotle. Straton wrote of using gears or cog-wheels, cranes, submarine navigation, and hoists with lead counterweights, and so on.[19]

Technology developed rapidly in the Hellenistic period when Greek science combined with the practical achievements of other civilizations, especially Egyptian. Alexandria had a library and museum. The Ptolemaic dynasty protected the city, and it became a cultural center. In the third century, Ktesibos, a barber's son, invented a water organ, a piston pump, a musical keyboard, metal spring, and water clock.[20] His student, Philo of Byzantium, wrote several works on mechanics describing pneumatic devices, catapults, techniques of siege warfare, mechanical toys, and a water wheel.

Archimedes (287–212 B.C.) was the greatest inventor of Syracuse. He was a great mathematician and physicist who discovered laws of hydrostatics, and of governing lifting devices. He worked on ship-building, built a planetarium that imitated the motions of celestial bodies, and developed military machines.[21]

People in the Hellenistic world, the Romans, Byzantines, and people in the

Arab world valued inventions. But, because of the influence and prestige of Greek science, these inventions were largely based upon scientific foundations. During the reign of Charlemagne, and after the Normans and Madziars invented the horseshoe, horse-collar, heavy-wheeled plough, water-mill, and windmill, some people of Medieval Europe dreamed of inventions that seemed fantastic at the time.

Toward the end of the Middle Ages, Roger Bacon wrote:

Ships could be constructed that move without rowers and can sail on rivers and on the ocean guided by one man at a greater speed that if they were full of rowers. Likewise vehicles that ride without any beasts pulling them and driven by unbelievable energy could be built, just as it is said that the chariots of the Ancients would ride armed with scythes. Flying machines could be built so that a man seated in the machine shall direct it by an ingenious mechanism and fly through the air like a bird. Likewise devices could be designed that although small could lift or press the greatest loads . . . .. Machines could also be constructed similar to those made at the command of Alexander the Great that could either sail on the surface of the water or dive beneath it.[22]

These daring and visionary designs are those of a Medieval friar, not of Leonardo da Vinci. While centuries would pass to make these ideas into reality, we can see that inventiveness is not unique to modern times. However, the modern cult of inventions and practical results has advanced to the point that Moderns (1) tend to disdain theoretical knowledge and pure science while (2) they excessively claim for themselves exclusive credit for many ideas and inventions they inherited from the past.

Nineteen

THE MATHEMATIZATION
OF SCIENTIFIC KNOWLEDGE

Another characteristic of the modern conception of science is the inclination to treat things as mathematical problems. To be scientific, we must reformulate everything in the language of numbers. The scientist no longer sees any reason to speak of qualities, substances, or causes.

The tendency to mathematicization is not really new. The Pythagoreans had a mathematical conception of science that would reappear in Plato and neo-Platonism. The Pythagoreans thought that number is the basic, indeed, the only, category of being. Mathematics, therefore, must be the science that deals with reality.

Plato treated mathematics as one of the sciences, but he thought that dialectics was higher than mathematics. Plato's Academy reportedly had an inscription over the door: "He who does not know mathematics may not enter." This witnesses to the role of mathematics in Plato's philosophy.

Dialectics concerned ideas that do not belong to the material world, while the object of mathematics was the sphere of numbers that reside between the ideas and the material world. In any case, mathematics had a crucial role in our knowledge of reality because Plato's cosmology taught that the sensible world arose by the application of mathematics.

Plato wrote in the *Timaeus* "[W]hen God began to put together the Heaven of the world, he used for this fire and earth. But it is impossible for two elements to form a beautiful structure without a third. There must be some link between them to join them. The most beautiful link is that which makes as far as possible one being with the things it joins. Mathematical proportion most beautifully achieves this effect." [1]

Plato later said that the basic elements are joined by mathematical proportions and that the elements themselves are constructed mathematically: "It is clear to everyone that fire, earth, water, and air are bodies. But a body by its nature also has volume and every volume must have a surface; finally, every linear surface is composed of triangles." [2] And he maintained, "It becomes clear to us that the four elements mentioned always come into being together. This opinion was not precise. In reality the four kinds arise essentially from the triangles of which we were just speaking; but three of them arise from the same triangle which has unequal sides and only the fourth is made from an isosceles triangle." [3]

Plato's vision of how the world arose is based on his conviction that the Demiurge used mathematical operations to shape indefinite and shapeless matter (*chora*).[4] Number causes the material world to reflect the World of Ideas.

The neo-Platonists accepted Plato's views. They attached great weight to the mathematical sciences alongside their interests in dialectics, religious questions, and mysticism. They thought of mathematics as the key to understanding how the structure of the material world is built.[5]

The mathematicization of the model of scientific knowledge arose from the earlier metaphysical assumption that material being has a quantitative structure. Aristotle regarded quantity as only one of the categories of being and as an accident that could not exist on its own. In Pythagoreanism and Platonism quantity had the status of a major category or the only category. In Platonism metaphysical reasons existed for thinking that our knowledge of the material world must be mathematical.

The scientific revolution of the sixteenth and seventeenth centuries was in large measure a continuation of Pythagorean and Platonic ideas. Nevertheless, the Renaissance did not forget the Middle Ages. Robert Grosseteste and Roger Bacon had led a strong mathematical movement during the thirteenth century. And Oxford University developed a mathematical conception of science. However, mathematics did not provide a complete knowledge of the world because it could not answer questions about the nature of things and the causes of nature.

Grosseteste was a neo-Platonist and ascribed a major ontological role to light. He said that light is the form of material bodies and the efficient cause of motion. In the beginning God created prime matter and light (*lux*). As light shines forth it produces space and particular beings. It then becomes the form of these beings.[6] Optics had to play a special role in this vision of being, and geometry was central to optics. Visible light (*lumen*) is only one aspect of metaphysical light (*lux*).[7] The laws of optics became the foundations for explaining the laws of nature.[8] When we use mathematics to describe and correlate physical phenomena, it provides us with no knowledge of causes.[9] Grosseteste parted company with the predominant use of mathematics in science. He adhered to the traditional explicit priority of metaphysics over mathematics and the physical sciences.

We are left with the question: what was the reason for the banishment of metaphysical thinking from scientific reasoning, for the exclusion of metaphysics from the order of the sciences, and for the complete mathematicization of the conception of science?

This appears to have happened under the influence of nominalism. Apparently, nominalism had caused the elimination of metaphysics. The elimination of universal concepts and the essences of things to which these concepts correspond shook the classical conception of science to its foundations. Nominalism's influence completely nullified what had been the core of metaphysical thought for two millennia. An empty word—*flatus vocis*—was all that remained of essences, causes, and the concepts that corresponded to them. Only then could the mathematical conception of a science liberated from metaphysics fully glory in its triumph. The modern conception of science based on the model of mathematical knowledge has two chief proximate sources: (1) a continuation of Pythagorean and Platonic thought; second, (2) the nominalism of the late Middle Ages.

Twenty

THE INFLUENCE OF NOMINALISM

The Greek conception of science was based upon the discovery that an object or object's aspect possesses universality (*to katholou*) . Concrete things must always be changing and accidental. So, our knowledge of concrete things must also be changing and accidental. This type of knowledge is opinion, not science.

Plato had a different understanding of generality than Aristotle. Plato associated generality with ideas that exist separate from the knower. The basis of generality is that many concrete things correspond to one idea. Many similar concrete things exist because they were produced according to the model of one idea in which they participate. Generality appears in relation to many concrete things because if we consider an idea in itself it is not general.

Aristotle associated generality with our mode of knowledge, not with separate ideas. Aristotle did not think ideas exist apart in a separate dimension of being. He maintained that a form exists in each thing. And this form is the foundation for a general concept. We express our general mode of knowledge in concepts, and concepts in Aristotle's teaching correspond to ideas in Plato's philosophy.

While Aristotle linked concepts to our human way of knowing by induction and abstraction, they have an objective basis in things. Form is the basis of concepts. The form allows a concept to appear because that form is what is definite and stable in a thing, while the matter is the source of indefiniteness and instability.

The Greek conception of science is ultimately based upon metaphysics. The structure of being makes a scientific and theoretical approach to reality possible. With the rise of nominalism changes arose in metaphysics and the theory of knowledge. The changes in the theory of knowledge stemmed from a change in the Western conception of a human being. The effects of nominalism influenced the conception of science and all areas of culture: ethics, art, and theology. Nominalism strongly shaped the character of the Renaissance and the Enlightenment, and its influences remain today.

Nominalism as such is no distinct school of philosophy. It is a general approach that left its mark in many areas. In the case of the conception of science nominalism undermined science as *theoria*. In its different forms nominalism denies the existence of forms of things and any objective counterpart to general concepts. Nominalism also undermines concepts as they associate with our human way of knowing things. Nominalism implies that, if we want to retain the conception of science and knowing that the Greeks transmitted to us, nothing, and no way, exists to know.

At this point we have two possible courses. We must: (1) abandon science or (2) change our concept of science. If we abandon science then we may supplant it with some form of irrationalism, mysticism, ecstasy, or even some state of love. If our conception of science changes, the new conception will depart from the Greek ideal and will move toward pragmatism and utilitarianism.

Nominalism permeates Western philosophy's history. It made its appearance first among the Sophists, then the Stoics, and in full force as a theoretical problem in the Middle Ages when it bore fruit in the controversy over universals. Thirteenth-century thinkers kept nominalism at bay. But it prevailed in the fourteenth century. With no exaggeration I must say that, thereafter, the more important schools of modern and contemporary philosophy have been based on the premises of nominalism. Consequently, *theoria* has had no place in the modern and contemporary conceptions of science.

Porphyry's *Isagoge*, an introduction to Aristotle's *Categories*, served as the starting point in the Medieval controversy about universals. Porphyry was a student of Plotinus. The *Categories* was one of Aristotle's early works. In it he discusses the main modes of being, the categories, more from a grammatical than an ontological viewpoint.[1] Porphyry's commentary thus presented the problem of the categories from a neo-Platonic and logical perspective.

The Medieval controversy about universals started in the eleventh century before Latin scholars knew Aristotle's *Metaphysics*. The controversy had metaphysical repercussions. But since scholars did not know Aristotle's metaphysics, they approached problems with Platonic idealism and nominalistic grammar. In the thirteenth century, nominalism moved beyond the confines of logic and grammar and found a metaphysical foundation in the teaching of John Duns Scotus. The metaphysics of Scotus would be more influential than that of St. Thomas Aquinas in the subsequent history of philosophy.

At the start of the *Isagoge* Porphyry writes: "Regarding species and genera, to begin with this, I will not say anything about whether they are something real or whether they are corporeal or incorporeal, and whether they occur separately or in sensible things, for it is too difficult a problem and requires extensive investigations."[2] This passage led to a tempestuous discussion that lasted centuries. The opposing parties were more firmly entrenched in their positions than if they were political adversaries.

The way Porphyry presented the problem was an important element in the controversy. Porphyry posed alternative solutions in such a way that, if philosophers rejected Platonism they had to fall into the trap of nominalism in some form. If we do not conceive species and genera following Plato, as something real, then they must be mental concepts, a position contemporary philosophers generally call conceptualism. If concepts have no reference to the ontological structure of things, they lose their essential function in knowledge. If they are not concepts of things, they are entirely subjective. Conceptualism must lead to nominalism. Names are universal only in their extension but have no content or meaning and we cannot speak of any essence in a thing.

Nominalism has far-reaching consequences. We lose intellectual contact with reality because the intellect has no real object. Are we then left only with sense knowledge, or does the intellect use some cognitive prosthesis? Nominalism must lead to a change in our picture of the structure of intellectual knowledge. And this will be an immanent cause of a change in our conception of scientific knowledge. What must we know, and how should we know it to know scientifically, if we have disabled one of the most basic functions of our intellect? Logic can provide no adequate answer.

Porphyry presented the problem badly. He provoked logicians to do metaphysics, and their metaphysical venture concluded in nominalism. If we ask about the ontological status of species and genera, whether they are real, should we not first answer the question—what does to be real mean? We cannot resolve this question on the level of grammar or logic. The question requires metaphysics.

Apart from some indirect fragments most often Platonic in character, twelfth-century philosophers knew no metaphysics. Yet their answers had metaphysical and epistemological consequences. When they started to philosophize on the level of two-fold abstraction about species and genera without having first addressed the question of why being is being, their image of the world and their conception of philosophy and science became distorted. This would affect other areas of culture.

The problems of nominalism and universals belong to metaphysics. Metaphysics, not logic physical science, or neo-Platonism properly analyzes them. But neo-Platonism was the only metaphysics philosophers knew in the late eleventh and twelfth centuries. They left Platonism and fell into the trap of nominalism.

The teaching of John Duns Scotus relates to nominalism. While Scotus spoke of concepts and nature, both of which nominalism rejects, he stripped his concept of being of all determination except non-contradiction.[3] Being as being appeared as univocal and without content. All later determinations would have to be artificial additions because they were not in the nature of being. If no content corresponds to the concept of being, then all the more the content of being cannot correspond to any categorical nature.

Scotus divorced being from nature, and nature from the concept in its function as a transparent medium. In this way Scotus' metaphysics became a basic point of reference for different forms of nominalism. Although Scotus speaks of nature, his concept of being is nominalistic because this concept has no content. Consequently, species and genera are without content. Scotus's metaphysics is a nominalistic metaphysics.

The departure from the essence of a thing as the object of theoretical knowledge did not cause science to disappear. But it bore fruit in a new attitude in science. Experience and experiment came to the foreground. This is especially evident in the views of William of Ockham. If things have no stable and universal essences that satisfy the intellect, then, to make them intellectually satisfying, we must subject things to investigations and experiments.[4] Instead of contemplating things, the scientist performs elaborate procedures to gather knowledge. While his or her short-term goal may be to confirm a hypothesis, in a longer view, the scientist is no longer aiming at *theoria*, but at another area of culture: utility.

Nominalism also strikes at the principle of causality and our knowledge of things through causes. Causality's roots are in the structure of being. Heteronomic elements that have no existence on their own being composed being.[5] A nominalist holds that simple and independent elements exist that we grasp in a sensory and intellectual intuition. Supposedly, these elements are self-evident and require no further explanation.

Consequently, a nominalist uses the category of sequence, not causality. Elements follow one another, but are not causally connected. The prior is not the cause of the posterior. A succession or sequence of events exists unconnected by any cause. Causality is ultimately based upon substructural compositions in being. But nominalism eliminates these compositions. Hence, causality can have no role in our scientific knowledge of the world.

Nominalism also rejects abstraction as the process that leads to the appearance of the universal. Abstraction is impossible because no immaterial content exists in things. In such a case what could be the role of the intellect? The intellect will grasp concrete things in an intellectual intuition or will perform operations upon signs.

Semantic operations have an authentic "career" in the history of science. Intellectual operations upon signs such as a nominalist understands signs do not require that signs have any meaning because nominalism eliminates meaning as a connection between the intellect and an immaterial aspect of reality. The sign has a new function: operability.[6] Crucial is not that the sign should have a meaning, but that we can subject the sign to operations. A sign is no longer a stable content expressed in a definition. It is something upon which we can perform specific operations.

When we base science upon nominalistic principles, it tends to focus upon establishing rules for working with signs. If we can reduce the meaning of a sign to its operability, we must conceive this operability in some framework of rules. This kind of science consists in rules for operations on signs.

The clearest paradigms for such operations are paradigms in logic and mathematics. In logic and mathematics we work with signs that are very poor in content or have no content. We can perform a wide range of operations upon them. Whether some act of knowledge is scientific depends upon whether we can perform such operations.

Under the influence of nominalism the same approach spreads to the so-called "real sciences" and the humanities. Logical and mathematical operability becomes the paradigm of scientific knowledge. Then, in science's name, we eliminate the particular sciences concerned with the contents or essences of things and classical philosophy from the sciences. Positivism would complete this grand task.

Nominalism introduces the category of the intelligibility of the concrete thing as such. This is diametrically opposed to the Greek conception of science.

For the Greeks, at best, the concrete thing may be an object of opinion, but never an object of science. Nominalist science rejects universality and focuses upon the concrete thing. How can we know the concrete thing scientifically? Will not this knowledge of the concrete merely be a changing opinion?

The idea that knowledge of the concrete thing can be science presumes the discovery of fixed rules and laws governing operations on concrete things. The relations expressed in these rules bear the whole burden of science and replace essences. Quantitative relations best express these relations because a number as such is a relation between different quantities. Part of the essence of number is relativity based upon specific operations or models. This is why nominalism worked with mathematics so well. Experience of concrete things in a quantitative aspect provides data that operations transform based upon specific laws. This is the essential dimension of science in a nominalistic sense.

Nominalism is a negation of Platonism: the universal does not exist; only the concrete thing exists. If only concrete things exist, what can we do with names that are universal? According to Plato universality is a property of ideas; and concepts or meanings are the subjective counterparts of ideas. Nominalism completely rejects concepts; or it allows them to remain, they but with no counterpart in things.

Conceptualism is the second position. It differs from Aristotle's position because it rejects any objective basis for concepts. Just as Plato's realism is the antithesis of nominalism, so Aristotle's realism is the antithesis of conceptualism.

Nominalism influenced theology, and, through theology, science. William of Ockham wrote about God's omnipotence: if God is truly omnipotent nothing can limit Him. The theologian must reject the existence of natures or essences because they would place limits on God's omnipotence. Since the Aristotelian doctrine that necessity govern scientific proof finds its is basis in natures or forms, this type of proof must collapse if we reject the existence of natures.[7] The result of these theological speculations was that content-based syllogistics lost its scientific value. Nominalistic assumptions started to guide the way theologians thought about God's essence. Subsequent theological voluntarism led to the destruction of the Aristotelian conception of science.

Would science then completely collapse? No. But a new conception of science arose based on syntactic relations expressed by rules. The new conception of science drew upon religious premises. It acquired a religious prestige and the old conception was occasionally condemned for theological reasons. The controversy about science expanded beyond science to religion. But the kind of theology in whose name the classical conception of science was condemned was based upon nominalistic principles and not on any biblical passages.

In different ways nominalism contributed to a change in the conception of scientific thought. Quantity, operations, and rules became central in science. Simultaneously, nominalism was able to conceal its metaphysical, epistemological, and theological assumptions and so protect itself from direct criticism. The strategy was cunning. Nominalism disavowed the field of scientific knowledge that could properly refute its principles. The science most capable of refuting nominalism is metaphysics. No surprise that nominalism attacked metaphysics most strongly, or that it continues to do so.

Part Five

SCIENCE AS TECHNOLOGY

Nicolaus Copernicus (1473–1543), Warsaw

Twenty-One

THE QUEST FOR AN EARTHLY PARADISE

The practical fruits of science and inventions that make life easier and express human ingenuity were a normal part of Ancient and Medieval life. In modern times we find a cult of science as something primarily useful, whose aim is knowledge's practical benefits and no longer truth for its own sake. Modern thinkers tend to disparage purely theoretical speculations as futile and fruitless.

Is this re-Easternization in science's purpose based upon ordinary practicality or upon something deeper, outside of science as such? The context of this change was quite complex and obscure. It included speculations on *Sacred Scripture* occasioned by the Protestant Reformation, Renaissance Hermeticism, and the spread of Cabalism. The Reformation, Hermeticism, and Cabalism combined into a single powerful current that changed the face of Western civilization. Too often we view the changes brought by the Reformation as limited to the field of religion, while we disregard or treat Hermeticism and Cabalism as marginal phenomena.

The Reformation affected more than the Christian religion, which split into Catholicism and Protestantism, and Protestantism into countless sects. Western civilization as a whole underwent important transformations: new conceptions of religion, art, morality, and the sciences. Fideism prevailed in religion. Fine art rejected representation. Morality rejected a final human end. In so doing, modern morality increasingly became entirely a matter of supernatural grace. Utility became science's new purpose.

Protestant thinkers contrasted Ancient Jewish culture to the "backwardness" of the Roman Empire and Catholicism. As Charles Webster notes, "The sparkling image of the agricultural and technological proficiency of the Ancient Hebrew culture was contrasted with the subsequent backwardness of the Roman Empire and the Catholic Christendom." [1] Protestant culture changed the face of Western civilization. It embarked on the reconstruction of Paradise on Earth upon new foundations, one of which was a modified science.

In Protestant interpretation the biblical account of the Earthly Paradise implies that before original sin, human beings and society were perfect and nature obeyed us. While the Fall essentially corrupted us, religion could heal us of sin and science could help us master nature. In the conclusion of *Novum organum*, a work that would change the face of science and the world, Francis Bacon wrote: "By the fall man lost the state of innocence and at the same time was moved down from a royal throne to the rank of creatures. Both consequences can be corrected in some measure in this life: the first by religion and faith, the second by skill and science." [2]

Millenarianism supported this new conception of science's role. Some Protestants, especially the English Puritans, accepted millenarianism. They thought that a war with the Catholic forces of the Anti-Christ must precede the establishment of a thousand-year reign of God on Earth. The New Kingdom, the New Paradise of Earth, or the New Jerusalem would be an era of science as predicted by the prophecy of *Daniel*. These Protestants saw their task to be to recover the lost learning:

> But at that time shall Michael rise up, the great prince, who stands for the children of thy people: and a time shall come, such as never was from the time that nations began, even until that time. And at that time shall your people be saved, every one that shall be found written in the book. And many of those that sleep in the dust of the Earth, shall awake: some unto life everlasting, and others unto reproach, to see it always. But they that are learned, shall shine as the brightness of the firmament: and they that instruct many to justice, as stars for all eternity.[3]

Protestant thinkers frequently cited this passage to support the idea that science should serve practical ends in the Kingdom of God on Earth. And it became an important part of the Protestant Reformation.

> Upon examination of the prophetic texts, it was immediately appreciated that the advancement of learning and the return of the dominion of man over nature were important elements in the eschatological scheme. In this context, the fourth verse of *Daniel* 12 was of crucial significance and it will be seen below that this verse became the common denominator of the puritan millennial forecasts. . ... Thus the advancement of learning became an important dimension of the general millennial scheme and by this means, science medicine and technology were assured of an integral part in the mentality of the English Puritans throughout the revolutionary decades.[4]

Many Protestants saw themselves as the ones who would unseal the books of *Daniel*.

These Protestant authors had far-reaching aspirations. They looked beyond the immediate practical benefits of using physical science to gain influence over the entire Earth and the entire cosmos. John Milton (1608–1674) wrote of the our mission:

> And finally the universal learning is complete. Man's spirit no longer limited to this dark home and prison will reach far and wide until it fills the whole world and all space with his divine greatness. At the end the magnitude of circumstances and changes in the world will be perceived so rapidly that for him who will dwell in the fortress of wisdom rarely will anything be able to occur in this life that is unforeseeable or accidental. He will truly be one to whose directions and domination the stars will be obedient. The Earth and sea will obey him. The winds and storms will serve him. Finally Mother

Nature will submit to him as if a god abdicated the throne of the world and entrusted his privileges, rights, and administration to him [man] as an advisor.[5]

The vision of future happiness of Earth drew on religious eschatological speculations based on the *Bible* and on Greek myths of the golden age.

The ultimate inspiration for puritan ideas on the Kingdom of God was of course biblical. The opening chapters of *Genesis* provided a model of the utopian state of innocence before the Fall; the *Pentateuch* and Historical Books gave an insight into the intellectual capacities of such patriarchs and rulers as Moses and Solomon; the Prophetic Books provided assurances of the return to an age of bliss. This later prophecy was confidently reiterated by certain Fathers, like Papias and Irenaeus, who embellished the account of the future Golden Age with materials drawn from Hellenic sources.[6]

Authors of different sects speculated on the exact date when the long-awaited Kingdom would begin. They frequently mentioned dates between 1641 and 1666.[7] The determination and fervor of the leaders of the new time came from their belief that the God of Israel sanctioned their mission. Rome's fall and the conversion of the Jews were supposed to preced the advent of the New Kingdom. The first was an expression of anti-Catholic sentiment. The second lay at the basis of philosemitism.[8] Adam and other biblical figures such as Moses, Solomon, Cain, Bezaleel, and Elias became quite energetic and active persons who had unique intellectual abilities and a plentitude of knowledge that empowered them to master nature.

Speculative reconstructions of the lives of Adam's progeny amplified the picture of Ancient technology and agriculture. Most of the patriarchs were thought to have played some part in the development of the crafts. For instance Tubal-Cain, son of Lamech, was "instructor of every artificer in brass and iron." . . . The secrets of medicine and transmutation had diffused to the Egyptians, before being fully restored to the Jewish race by Moses. . . . From Moses this sacred wisdom was imparted to Bezaleel and Aholiab, who were supposed to have possessed "all manner of workmanship, as well as knowledge of the general principles of natural philosophy. . .." Ultimately divine wisdom was vested in Solomon, whose empire and wealth were established on technological and scientific foundations. The Song of Solomon presented an optimistic account of the potentialities of the fallen man.[9]

The approaching Kingdom of God would be the Paradise restored or the New Jerusalem.[10] In Joachim of Fiore's terminology, this period was also called the Age of the Holy Spirit, for nothing would be hidden any longer from human beings. We would regain Adam's knowledge of the world and would be able to rule the world as did Adam.

Thus Adam was granted dominion over all creatures, but only on condition that he displayed a lively scientific curiosity about his environment. Intellectually Adam represented an ideal balance between the active and the contemplative life. His knowledge was extensive, indeed encyclopedic, as witnessed by his naming of creatures (*Genesis* 1: 28), which was customarily regarded as including an ability to decipher the nature of every creature from its name. Man's return to a state of grace would be rewarded by a restitution of this dominion over nature.[11]

Adam as he was before his sin became the prototype for a new human being who would be endowed with many practical skills based on knowledge.[12]

To realize these aspirations science had to be redirected to practical ends. The conception of science and its place in society changed. Religion sanctioned the change. It was no mere result of a methodological change within science. The change was accompanied by a hope and awareness comparable to what was evoked by Christopher Columbus's discoveries. People expected better times because of science.[13]

When the new conception of science matured in the mid-seventeenth century, the influence of Protestantism was becoming weaker, and European culture started moving into a phase of far advanced secularization. Herbert Butterfield thinks that the establishment of the new conception of science in Europe was an event as significant as the appearance of Christianity. At the same time, this new type of scientific civilization could be separated from Christianity and from the heritage of Greece and Rome, and the center of civilization moved in the late seventeenth century from the Mediterranean Basin to the English Channel.[14] The conception of science would purify itself of explicit religious references but science's mission would remain the same—it would be the chief tool for transforming human beings, society, and the world.

Twenty-Two

THE END OF KNOWLEDGE:
COMPLETE UTILITARIANISM

In the Greek world knowledge for knowledge's sake (*theoria*) had scientific, cultural, and social value. The disinterested quest for truth was science's primary end. Science for the sake of truth held the highest position in culture. And wise people held the highest position in society. The Greeks treated the possible practical applications of their scientific discoveries with some contempt. They thought that practical occupations involving physical exertion were not fitting for a free human being. For example, because Greek culture chiefly aimed at making us spiritual, Aristotle maintained:

> There can be no doubt that children should be taught those useful things which are really necessary, but not all useful things; for occupations are divided into liberal and illiberal; and to young children should be imparted only such kinds of knowledge as will be useful to them without vulgarizing them. And any occupation, art, or science, which makes the body or soul or mind of the freeman less fit for the practice or exercise of virtue, is vulgar; wherefore we call those arts vulgar which tend to deform the body, and likewise all paid employments, for they absorb and degrade the mind.[1]

The Romans inherited much from the Greeks, but they considered politics, not science, the high point of a free human being's action. Hence, Marcus Tullius Cicero said:

> I suppose that he who wanted and was able to achieve one and the other, that is, to know both the statutes of the ancestors and the precepts of philosophy, has gained everything that composes glory. If we had to make a wise choice between these two ways, then, although the peaceful style of life consisting in dedication to the most sublime sciences and arts may seem to be happier, yet we must maintain that a life dedicated to the state is certainly more praiseworthy and holy.[2]

Christianity had a new respect for human work: because of original sin work was part of our human lot.[3] Nevertheless, theoretical science still held quite a high position in the hierarchy of the sciences and culture.

This changed during the Renaissance when the mechanical arts and practical skills became greatly esteemed and purely speculative knowledge came under attack. The Renaissance brought a complete transformation of values in scientific culture due partly to men of science and partly to religious activists, primarily those associated with the Reformation.

In *De tradendis disciplinis* [*On Spreading Learning*], the Renaissance, Spanish, humanist Juan Luis Vives encouraged learned men to study techniques of building, navigation, manufacture of textiles, and the secrets of industries.[4] François Rabelais thought that a complete education should include knowledge of different industries. George Agricola emphasized the importance of metallurgy. Humphrey Gilbert developed a new program of education that would produce theoreticians and colonists with a wide range of practical skills. Thomas Starkey said that a human being's perfection consists in possessing knowledge that produces practical benefits and can be useful to others.[5] Quite a large group of sixteenth-century scholars accepted his rehabilitation of industry and practical arts. Their acceptance of it was a factor in the change in the conception of science and in society and civilization.

New geographical discoveries that presented new challenges to many European nations surely strengthened this practical attitude. In hope of material gain European rulers promoted initiative and industry in response to these discoveries. Colonization helped cause redirection of science to practical ends to become a need of the moment.

In religious terms, the practical attitude became increasingly endorsed and the theoretical attitude increasingly rejected. Religion entered the controversy about the conception of science. A religious fanaticism led to intellectual changes and to spectacular gestures such as the burning of Medieval books, including works of John Duns Scotus, in the courtyard of the New College at Oxford in 1553.[6]

Martin Luther frequently and severely criticized classical culture and the achievements of Scholasticism. He was especially fond of attacking the notion that the human reason should reach beyond practical matters. He wrote: "God gave us reason only so that it would rule here on Earth, that is, it possesses the power to create laws and decrees in such matters of life as drinking, eating, clothing, and in external discipline and honesty."[7]

If reason trespasses into the area of religion or theoretical knowledge it should be cursed. Luther maintained that reason is "God's worst enemy." A "diabolical prostitute . . . , it can only blaspheme and dishonor what God said and created."

Near the end of his life Luther spoke in a similar tone in a sermon in Wittenberg: "The reason is a diabolical prostitute. It is a prostitute, a real prostitute, a diabolical prostitute on fire with itches and leprosy who should be trampled underfoot and destroyed along with her wisdom. . . . Throw filth in her face to make her ugly. The reason is drowned and should be drowned in baptism. . . . The abomination deserves to be thrown into the filthiest corner of the house, into the latrine."[8]

Although Luther used extreme language, ultimately he based his view upon the idea that the effects of original sin are so extensive that human reason cannot perform its normal functions and cannot know the truth. Baptism and grace do not repair this damage. Therefore, theoretical science cannot reveal truth. It reveals errors. (*"Omnes scientiae speculativae non sunt verae . . . scientiae, sed errores"*).[9]

Protestant theology of original sin underlies the rejection of theoretical knowledge in modern times. The functions of the theoretical reason and the theoretical sciences were branded as having come from the same contaminated source. All reason's energy was redirected to practical ends because, somehow, reason was not totally corrupted in practical matters. Human beings had been cast out of Paradise, and we could recover Paradise in this way: by making the Earth once again subject to us and taking command over all nature.

The change in direction for reason would mark a return to our original practical skills. Another practical end redirected science. And *theoria* metamorphized into technology.

Protestant utilitarianism had deep roots in religious ideas and changed the direction of the development of Western civilization. Science was affected since it would be the primary instrument for changing and mastering the world. Philosophy easily absorbed Protestant ideas. At the time, many philosophers, including Francis Bacon, were Protestants.

Bacon discussed the problem of defining science's purpose on many occasions. In his writings he firmly condemned any end for science apart from utility. In his "Foreword" to the *Great Instauration*, he wrote: "Finally we want to remind everyone to keep in mind the true ends of science and not to pursue science for satisfaction or pleasure, or to be able to look down on others, or for profit, or for glory or power, or any other lower ends of this kind, but for the sake and use of life and to perfect it in love and guide it."

In his *Great Instauration* he added: "Lastly, I would address one general admonition to all; that they consider what are the true ends of knowledge, and that they seek it not either for pleasure of the mind or for contention, or for superiority of others, or for profit, or fame, or power, or of any of these inferior things; but for the benefit and use of life; and that they perfect and govern it in charity." [10]

In these passages Bacon mentioned different false ends of science (such as satisfaction, pleasure, profit, fame, and power). But he made no mention of the primary motivation in Greek philosophy: knowledge for knowledge's sake, or for truth. And he did not mention knowledge for knowledge's sake when he presented the proper end of science.

In another passage Bacon said:

For our science has as its end not to find arguments and to find arts, not to find propositions in agreement with first principles, but to find the first principles, not to find probable reasons but instructions and directions for carrying out works. A different effect will result from a different intention.

There we defeat and hold down our adversary in a dispute, here we defeat and hold down nature in our action.

Bacon said that the end he proposed for his science was discovery of (1) arts, (2) principles, and (3) designations and directions for works, not discovery of (1) arguments, (2) things in accordance with principles, and (3) probable reasons. Because his intention was different, so, too, he said, was "the effect of the one being to overcome an opponent in argument, of the other to command nature in action."[11]

Simultaneously, Bacon distanced himself from philosophy as dialectics or the art of argumentation from probable premises. For Bacon science's true end is to find rules that empower us to master nature, not propositions to defeat our adversaries in debate.

In his *Aphorisms* Bacon returned to the problem of knowledge's end. He considered the reasons why the old science made little progress. His conclusion was that Ancient science had the wrong end. "It is not possible to run a course aright when the goal itself has not been rightly placed. Now the true and lawful goal of the sciences is none other that this: that human life be endowed with new discoveries and powers."

In his *Novum organum* he added: "Again there is another great and powerful cause why the sciences have made but little progress; which is this. It is not possible to run a course aright when the goal itself has not been rightly place. Now the true and lawful goal of the sciences is not other than this: that human life be endowed with new discoveries and powers."[12]

Thus science progresses only when it aims at new discoveries and finding new "powers."

In aphorism 124 Bacon directly compared his notion of the proper end of science and the classical one. He recognized that he would face the objection that had "not touched upon the true and best term or end of science, for the contemplation of truth is a thing worthier and loftier than all utility and magnitude of works."

He gave a classical argument to support the objection: "A long and restless sojourn with experience, matter, and fluctuating particular things fastens the mind, as it were, to the Earth, or rather casts it into an abyss of confusion and draws it away from the serenity and peace of abstract wisdom, a state much more divine."

Bacon said he gladly shared this view. He claimed that he primarily had in view contemplation of truth: "[A] true picture of the world, a picture such as we find it and not such as someone makes according to his own reason."

Bacon maintained, however, that we cannot form a true picture of the world "without making a very precise section and anatomy of the world." He claimed that human imagination had created in philosophical systems "absurd models and apish likenesses of the world that should be "completely removed."

He maintained that we should be aware how great a difference exists between the idols of the human mind and the ideas of God's mind. He said the first are "arbitrary abstractions", while the second "are true signs of the Creator in creatures—impressed as he impresses and describes them in matter by true and exactly drawn lines." From the magnitude of difference between our ideas and God's ideas, Bacon concluded, "Truth therefore and utility are the very same things, and works themselves are of greater value as pledges of the truth than as contributing to the comforts of life."

Bacon said further:

Again, it will be thought, no doubt, that the goal and mark of knowledge which I myself have set up . . . is not the true of the best; for the contemplation of the truth is a thing worthier and loftier than all utility and magnitude of works. . . . Truth therefore and utility are here the very same thing; and works themselves are of greater value as pledges of truth than as contributing to the comforts of life.

In the original Latin text, the crucial phrase is: "*itaque ipsissimae res sunt (in hoc genere) veritas et utilitas.*"[13]

In these passage Bacon tried to (1) reconcile utility with truth and (2) show that truth needs utility. Knowledge based on *a priori* or fantastic constructions of the mind, such as philosophical systems, is not true if we do not force it to confront reality. The way of "anatomy and section" allows us to draw a true picture of the world. This anatomy and section occurs when we strive for utility. Therefore, truth and utility need each other, are identical.

The attentive reader will notice a flaw in Bacon's eloquent argument. "Anatomy and section" allow us to see only a partial aspect of reality, that which appears when we strive for utility. But are useful beings all that exist in reality? Is utility reality's most important aspect? Doubtful that this is so.

Subordination of truth to utility distorts our image of reality by reducing reality to one of its less important aspects. Philosophical knowledge of reality does not need to be a fantasy. It is chiefly an approach to knowing reality under other aspects more important than mere utility. Other criteria than practical experience can exist for determining whether our knowledge agrees with reality. If practical experience were truth's only criterion, and if this experience had to be limited to the aspect of utility, then the philosophical image of the world based on this criterion would essentially distort our vision of reality. In this case utilitarianism would become an *a priori* principle, or in Bacon's words, an "idol" that distorts our philosophical knowledge of the world. Bacon's view of the order of knowledge marks an essential change with many consequences for culture and civilization.

When we confound truth with utility in philosophy, we change our picture of the world and of our place in it. We popularly know Bacon's works by abbreviated titles, but their full titles testify to this. The full titles provide food for thought, for example: *Novum Organum. Aphorisms on the Interpretation of Nature and the Kingdom of Man; The Greatest Birth of Time or the Great Restoration of Human Rule over the Universe* (*Temporis partus maximus sive instauratio magna imperii humani in universum*).

Twenty-Three

THE INFLUENCE OF THE EAST

1. Hermeticism

In 1462 or 1463, Cosimo de'Medici sent a manuscript of the *Corpus Hermeticum* that he had received from the monk Leonardo da Pistoia to Marsilio Ficiino, founder of the Platonic Academy in Florence. Ficino translated the work, and published it in 1471. By the end of the century 16 editions of this book existed, not counting translations and works that developed its contents.[1] This clearly shows that Hermeticism evoked great interest among Renaissance intellectuals and strongly influenced them.

The *Corpus Hermeticum* was consisted of two groups of writings: (1) The *"Asclepius"* or *"De voluntate divina"* ("On Divine Will") and (2) the *"Pimander"* or *"De sapientia et potestate Dei"* ("On God's Wisdom and Power"), and the *"Definitiones Asclepii,"* which consisted of 15 short dialogues written in Greek.[2]

Apart from the *Asclepius* and fragments of the dialogues, the *Corpus Hermeticum* had been unknown in the West. Until the early seventeenth century, most followers of this tradition thought that Hermes Trismegistus, whom Plato identified with the Egyptian god and king Thoth, was the author of the *Corpus Hermeticum*. Hence, the work bore the name "Hermes." They also thought that Hermes was the original source of the teachings of Egyptian, Jewish, and Greek wise men.

Other followers of this tradition have occasionally regarded Hermes Trismegistus as a more recent Moses. This tradition lists the following as Ancient theologians: Adam, Enoch, Abraham, Noah, Zarathustra, Moses, Hermes Trismegistus, the Brahmans, the Druids, King David, Orpheus, Pythagoras, Plato, the Sybil, and the *New Testament*.

Roger Bacon presented a somewhat different genealogy. He also thought that Adam received philosophy from God. But, after the Fall, Zarathustra received philosophy. Then Prometheus, Atlas, Mercury, Hermes, Apollo ,and Asclepius. Philosophy's rebirth occurred in the time of Solomon and came to maturity in the time of Aristotle. Bacon thought that the Greeks were the students of the Hebrews and shared imperfectly in the Revelation of the Word, and they even recognized to some degree the mystery of the Trinity. Their opinion facilitated the spread of Hermeticism in Western thought and influenced the change in the conception of science.[3]

In 1614, Isaac Casaubon's laborious research showed that the *Corpus Hermeticum* had been written in the second or third century A.D.. So, it could not have been the original source of wisdom (*prisca sapientia*).[4] Nevertheless, the influence of Hermeticism remained strong in the seventeenth century, and many thinkers still respected the *Corpus Hermeticum*. Martin Bernal notes that Europe published more alchemical books during the 1750s than it did in the entire preceding century.[5]

From the temporal perspective, we can say that the *Corpus Hermeticum* was a Hellenistic amalgam of Platonism, Stoicism, Judaism, and Christianity presented in a framework of magic and gnosticism.[6] This syncretism gave the impression of a primordial body of wisdom containing the beginnings of different religions and philosophies. In reality, these common elements evince that Hermeticism derived from those religions and philosophies. It was not their ultimate source.

Researchers think that Hermeticism, an important influence on the change in the conception of science, infected most Renaissance intellectual schools, and especially the heralds of the new conception of science. In his many works, Francis A. Yates has played a major role in showing Hermeticism's influence upon Western thought.[7]

Hermeticism put high value on knowledge as power, knowledge that provided mastery of the forces of nature and empowered people to exploit those forces. André-Jean Festugière's four-volume work on Hermeticism shows that, in the Hellenistic world, especially in Egypt, the Aristotelian ideal of knowledge for the sake of knowledge declined and gave way to the ideal of knowledge for the sake of practical benefits.[8]

The *Corpus Hermeticum* maintained that the Egyptians knew how to make statues, which, by the infusion of the World Spirit, could move and speak.[9] In Alexandria the Egyptian god Thoth was identical with Hermes. Thoth was the patron god of magic arts and occult knowledge. He was the "Mysterious" and the "Unknown." Thoth had the role of the Demiurge in the production of the world. By his voice he had created things. He was also the patron of inventions and clever minor arts. The Alexandrians identified Thoth with Hermes because Hermes was the patron of similar arts. But they also identified him with Moses.[10]

Some ideas in Hermeticism had quite a broad appeal. Most people would like to live longer than the normal life span. Hermeticism promised that it was about to discover the secret for prolonging life. Most people would like to know the future. Hermeticism promised them this ability. Most people desire wealth. Hermeticism claimed to be on the verge of solving the mystery of transmuting other metals into gold. No wonder that the fantastic promises of Hermeticism attracted the masses, learned men, and statesmen alike.

Magic consists in the skilful manipulation of hidden relations of sympathy between material substances, demons, planets, and gods. Ancient magicians supposedly possessed secret knowledge of these relations and thought that their knowledge had come from the gods. To obtain the desired result magicians would use a specific terminology and utter different incantations.[11]

Empedocles was an Ancient Greek philosopher who came close to being a magician. His theory of the four elements (water, fire, earth, and air) and two forces (love and hatred) and of a primevel fire constitued part of a wider vision that proclaimed Empedocles a god.[12]

Reportedly, Empedocles promised to his adept Pausanius that he could acquire the power to rule the weather, and summon winds, storms, floods, and droughts. He also said that, by his occult knowledge and power, he could free people from their suffering, especially in old age, and could even give thim immortality. He maintained that his words had magical power.

Empedocles' student Gorgias witnessed the practice of magic. In Greek culture people regarded such aspirations as a scandalous expression of pride and impiety, and Hippocrates criticized them. Empedocles' flirtation with magic had a practical, not a theoretical, aim. Hence he would later play a great role in the formation of the foundations of Hellenic Hermeticism and Arab Sufism.[13]

During the Renaissance Ficino's work on magic, *De vita coelitus comparanda*, was quite influential. It was the third book in his *De vita libri tres* (1489).[14] By the mid-seventeenth century over 30 editions of it had appeared, a sign of the unfading interest in magic among intellectuals.[15]

The idea of hidden properties (*proprietates occultae*) appeared in occult magic. The expression came from medicine. Galen had introduced the Greek term *idiotes arretoi* to describe inexplicable and irrational forces that have the power to cause visible effects. Nineteenth-century thinkers coined the term "occultism" (from the Latin *occultus*) to designate trends that appeal to mysterious powers that science does not grasp.[16] Ficino sought the sources of these hidden properties in Heavenly bodies that influence the Earth by radiation (*radii*).[17]

Occult and demonic magic gradually changed into what he called "natural magic." But even when it was natural, it was still magic. Ficino did not hesitate to say that, with the help of natural objects, the magician, in this case the natural magician, controls the beneficial powers of natural bodies for medical purposes. Just as a farmer works in agriculture (the cultivation of fields, so the magician or priest works in "*mundi*-culture" (cultivation of the world) and uses knowledge to connect the higher beings of the cosmos with the lower beings.[18]

Cornelius Agrippa referred to a "natural magic" by which we can learn learning the properties (*virtutes*) of all natural and Heavenly bodies." He said that serious study of the order of these properties enables us to learn nature's "hidden and secret powers" by which nature connects lower and higher things "by their mutual application to each other." He further maintained that the person who made this art "a servant obtains incredible marvels not so much by art as by nature when he works upon these things." Hence, he said:

> [M]agicians are like cautious explorers of nature guided only by what nature has already prepared as they join active elements with passive elements, and they frequently anticipate effects that would normally be regarded as marvels but which in reality are only anticipations of natural operations.[19]

This conception of natural magic paved the way for a practical approach in the natural sciences. The natural sciences would aim at learning the laws and forces of nature in order to use them. From this tradition Francis Bacon borrowed the idea of science as power and as the servant of nature, and that science at the same time empowered human beings to dominate nature.

Pado Rossi notes this when he writes: "We have already seen that the metaphysical aspects of magic and alchemy had little or no influence on Bacon; but he did borrow from this tradition the idea of science as the servant of nature assisting its operation and, by stealth and cunning, forcing it to yield to man's domination; as well as the idea of knowledge as power."

Simultaneously, Rossi warns us against forming a picture of Hermeticism based upon Bacon's views, as did Yates. According to Rossi, Yates maintained that Bacon presented in a more up-to-date language aims and values characteristic of the Hermetic tradition. Supposedly Bacon maintained that we could never count magic and alchemy as sciences precisely because of the excessive importance they attach to individual authority of individuals and their over-hasty judgments. The rules that the procedures of magic and alchemy contain can never reach the level of a method, because we can never establish their codifiable character.

For this reason, they will always remain secret rules, formulated in a symbolic language having nothing to do with modern chemistry's symbolism and referring "through a series of analogies and correspondences, to the Whole, to Universal Spirit, to God. The alchemist cannot codify his method, nor make it a *public* knowledge, available for others and to be used by others."[20]

Besides magic, the Hermetic tradition contained alchemy and astrology. Alchemy aimed to transmute other metals into silver and gold. The alchemists assumed that every metal is composed of the same physical base but each metal has a different "soul." The soul in a metal was, to some degree, impure. And the alchemist needed a special procedure to obtain the purest soul that gold possesses. The alchemist used a laboratory to perform experiments in purifying the souls of metals. These experiments became the basis for the newly emerging experimental science.

Astrology's basis was the idea that an observable connection exists between human actions and the motions of the planets. This enabled the astrologer to foresee the future, but did not enable him to change it.[21] As technology provided increasingly precise scientific instruments, astronomy would fill the gaps in astrology.

The Eastern sources of Hermeticism influenced the institutional organization of scientific work. Eastern civilizations commonly restricted knowledge to a small caste of priests. Something similar took place among modern "Hermeticists," but without the trappings of a formal religion. Science was the work of an elite that worked together. And they did not intend their findings for the general public. Hence, in a letter to Johannes Trithemius, Agrippa cautions, "I advise you to observe diligently this precept: you may pass on general matters to the wider world, but you must pass on the higher mysteries only to the better circles and to selected friends."[22]

The East generally accepted idea that scientists should work together and organize into different functions and specializations in a scientific workshop was generally accepted in the East. But disagreements existed as to whether they should divulge their findings to the public.

We can understand these disagreements because power is one of knowledge's fruits. If scientific learning opens up countless possibilities for changing the world, should everyone have equal access to it? Should we not carefully guard a kind of knowledge that is power?

Many occult books exist in the Muslim world written in a special cryptic language. One that reached Europe was the pseudo-Aristotelian *Secretum secretorum* to which Roger Bacon wrote a commentary. This work supported the opinion that knowledge is not for everyone. Bacon stated that the secrets of the sciences are not inscribed on the hides of asses and rams.[23]

In modern times, when learned men such as Robert Boyle wanted to make their findings public, merchants, craftsmen, and statesmen opposed them. Merchants and craftsmen feared the loss of profits. Statesmen saw a potential threat to the security of the state.[24] While many people regard Francis Bacon as the champion of the democratization of science and making science a public matter, in his fictional account of the island of Bensalem, even the rulers did not know all there was to know about the work of the scientists.

According to Rossi, "Bacon proposed to the European culture and alternative view of science. For him science had a public, democratic, and collaborative character, individual efforts contributing to its general success." Rossi maintains further:

The men of science, in the New Atlantis, lived in solitude. Their place reminds us of a university campus cut off from the daily concerns of common mortals. But there is something else: the scientists of the New Atlantis held meetings to decide which of the discoveries that had been made should be communicated to the public at large and which should not. Some of the discoveries that they decided to keep secret were revealed to the state; others were kept hidden from political power. On the uses that might be made of scientific and technological discoveries he was no optimist.[25]

Today, in the Western democracies much technological research is a State or business secret. Knowledge provides physical and political power. And it is a great source of profit

The Hermetic approach to knowledge resulted from a world view in which human beings no longer stand in awe before the cosmos, worship God, or have as their purpose to behold God face to face in the beatific vision. Human beings actively take part in changing and making things and use their discovery of terrestrial, cosmic, and divine powers.[26] Francis Bacon repeated, after the Hermeticists, that only God and the angels have the right to be spectators in the stage of human life. Humans cannot be idle spectators.[27]

We should not allow Giovanni Pico della Mirandola's (1453–1494) use of the word "dignity," in his *Oration on the Dignity of Man*, mislead us. His vision was not the Christian one of a human being as a person. He thought of a human being as a magus or sorcerer who weds Earth with Heaven.[28]

A passage from this work shows how strongly Eastern thought influenced him: "Most esteemed Fathers, I have read in the ancient writings of the Arabians that Abdala the Saracen on being asked what, on this stage, so to say, of the world, seemed to him most evocative of wonder, replied that there was nothing to be seen more marvelous than man. And that celebrated exclamation of Hermes Trismegistus, 'What a great miracle is man, Asclepius' confirms this opinion."[29]

Significant is that Pico quotes Hermes Trismegistus and Asclepius because this citation shows how the Renaissance increasingly accepted Hermeticism and how syncretism could blend different religions together. Hermeticism was an Eastern syncretism. During the Renaissance Hermeticism thoroughly mixed with Christianity.

During the late sixteenth century, Hermeticism started to spread in Protestant lands. Not surprising since Protestantism and Hermeticism had similar aspirations: to dominate the Earth and the cosmos. P. M. Rattansi maintains, "Most of the motives which Merton regarded as consonance between Puritanism and science are, indeed, to be found intertwined in the development of Hermeticism in the Protestant lands from the late sixteenth century onwards."[30]

The Aristotelian approach and the modern approach to that Hermeticism inspired differed because of a change in the end of knowledge. This joined itself with an essential transfiguration of the conception of science: science in the Aristotelian sense was no longer necessary, and even ceased to be regarded as science. Simultaneously, the new "scientific" science had quite an obscure starting point: hidden powers (*proprietates occultae*), cosmic sympathies and antipathies, and reliance on divine illumination to gain knowledge of the world.

Reliance on divine illumination was a sign of pessimism regarding the human ability to gain knowledge, in contrast to the optimism of the Greeks in their pursuit of knowledge expressed in their desire to discover the causes and principles of being. The moderns regarded the attitude of the Greeks as a sign of pride.

Rattansi maintains: "Such a view promoted a sharp break with the Aristotelian assumption of the capacity of human reason to penetrate to first principles. If knowledge of nature was primarily the knowledge of occult virtues and secret sympathies and antipathies, it could come only from a revelation, from a blinding vision vouchsafed by a divinity."[31] The civilizations and scientific approaches of the Ancient and modern world parted ways and even became enemies.

2. The Cabala

Along with Protestantism and Hermeticism the Cabala also influenced the formation of the modern conception of science. We associate the Cabala with

Judaism. It means "received knowledge." The term first appeared in literature in the eleventh century. But the general framework of the Cabala as a doctrine took shape in the second and first century B.C. among the Essenes.

The Cabala is a body of esoteric, or mystical, knowledge concerning God and the universe reputedly revealed to chosen saints and transmitted thereafter to some chosen and privileged persons.[32] The conception of Cabalism has associa-tion with a passage in the prophecy of *Daniel* (12:10) that mentions concealed knowledge. Many Protestants, who thought they were the elect in the possession of the knowledge that would allow a return to Paradise on Earth, later cited this passage.

The most important passage is in the fourth book of *Esdras* (14: 5, 6). This states that, on Mount Sinai, Moses received the Law that he proclaimed to the people and a miraculous knowledge that he could not make public but should pass on only to selected individuals. (Only the first and second books of *Ezdras* became part of the Catholic canon of Sacred Scripture.)

The Cabala is supposedly the tradition of passing down this marvelous knowledge. It contrasts with the *Torah* as the body of universally accessible knowledge.

The Cabala developed with special intensity in Medieval Spain. It reached its high point in the thirteenth century when the Cabalistic book the *Zohar* appeared in 1275. In 1492, Spain expelled its Jewish population, and the Cabalistic movement relocated to new centers in Italy, France, Turkey, and England.[33] During the fourteenth century, Poland was one of the strongest centers of Cabalism. From Poland its influence spread to Germany.

The Cabala is divided into a theosophy (more theoretical in character), and a more practical theurgy. We may find in the Cabala some neo-Pythagorean and neo-Platonic influences. It presents a picture of reality based upon emanationism. The attributes of God (called "Sephirots") play an important role (*Gloria, Sapientia, Veritas, Bonitas, Potestas, Virtus, Eternitas, Splendor,* and *Funda-mentum*). (The term *Sephirot* comes from the terms *en-soph* [nameless]).[34] These attributes are also the divine powers by which God created the world.[35]

Cabalists attributed a magic power to the letters that formed the Tetragrammaton (YHWH, or "Yahweh"), the letters in the other names of God, and they even regarded the entire Hebrew alphabet of 22 letters as endowed with occult power. Some Cabalists thought they had the knowledge of how to join the letters by which God created Heaven and Earth. Numbers, like letters, and certain arrangements of numbers, had a magical and creative power because they are the language of God.

The modern Cabala started with Isaac Luria (1533–1572). The Cabala was already finding a place in Christian thought in the thirteenth century. Raymond Lull (1225–1315), the author of the renowned *Ars magna* (*Great Art*) drew extensively from the Cabala. He concentrated on the mysticism of letters and numbers, and developed a new method of scientific and philosophical knowledge. But he used Latin, not Hebrew, letters. Lull wanted to unite the three great monotheistic religions (Judaism, Islam, and Christianity) by showing their common principles. He also tried to find the foundations for all true knowledge. The common points were nine divine attributes (*Bonitas, Magnitudo, Eternitas, Potestas, Sapientia, Voluntas, Virtus, Veritas*, and *Gloria*), the Aristotelian theory of four elements (earth, air, fire, and water), and the so-called common *topoi* (difference, agreement, opposition, principle, means, end, greater, equal, and lesser). Lull wrote the letters that corresponded to the attributes (B, C, D, E, F, G, H, I, K) in three geometrical figures—a triangle, circle, and square. These figures, in turn, respectively symbolized the Holy Trinity, Heaven, and the four elements of the Earth. The letter "A" signified the first nameless attribute.[36] The result was a table where Lull could put together different combinations that would always provide religious, philosophical, or scientific truth.

Étienne Gilson maintained , "Lull's art largely consisted in begging ahead of time the principles from which the expected agreement must necessarily follow." Gilson claimed that the technical pedagogical processes by which Lull "believed he could teach the uninstructed and convince the unbelievers contained the germ of an idea which had quite a future." Gilson asserted that Lull's revolving tables on which he inscribed his fundamental concepts constituted the first attempt to develop the "combinative art" that Gottfried Wilhelm von Leibniz, who remembered Lull's work, later failed to constitute and which is, "by no means dead."[37] Gilson saw no role of the Cabala in the formulation of Lull's art.

In the late fifteenth century, Pico was in close contact with Jewish Cabalists in Spain. He regarded the Cabala as the sum of Jewish revelations. Most importantly, he thought that the Cabala confirmed Christian truth. He concluded that the Cabala could assist in converting the Jews. So, he want to spread it among Christians. Pico distinguished a combinatory art (*ars combinandi*) in the Cabala where principles existed for joining Hebrew letters, and an art concerning ways of acquiring higher powers.[38] This conception of the Cabala reflects Hermetic influence.

The Cabala became popular in Christian circles, which opened an avenue for Jewish influence on European culture. Because of Johannes Reuchlin (1455–1522), the Cabala strongly influenced the Reformation and was presented as an alternative to Scholasticism. Pico's student Reuchlin was the first author outside Judaism to write a systematic work on the Cabala, *De arte cabalistica* (1517).

Reuchlin wrote that, by divine illumination, the Cabala allows us properly to interpret letters, numbers, and the *Sacred Scriptures*. He thought that the letters of the Hebrew alphabet had a numeric value, that we could numerically express the names of God and the angels. And because the Cabala regarded the Hebrew alphabet as instrumental in the creation of the world, we could treat the Cabala as a mystical mathematics.[39]

Conversely, mathematics would thereby move to first position among all the sciences. In this way the mathematical picture of the world started to become increasingly dominant.

The Cabala influenced Martin Luther, Philip Melanchton, Heinrich Cornelius Agrippa von Nottesheim (1487–1535), and Francesco Zorzi (1465–1540). The Cabala played a special role in creating a new outlook on the natural sciences in the writings of authors such as Paracelsus (1494–1541), Hieronymus Cardanus (1501–1576), Johannes Baptista (1537–1644), and Robert Fludd (1574–1637). The Cabala influenced some mystics, such as Valentin Weigel (1533–1583) and Jacob Boehme (1575–1624).

Pico's idea that the Cabala could be helpful in converting the Jews became popular. So, translations of Cabalistic works spread widely among Christians. Joseph de Voisin (1610–1635), Athanasius Kircher (1602–1624), and Christian Knorr Baron von Rosenroth played important roles in disseminating Cabalistic works.

The Cabala also influenced the philosophical systems of Gottfriend Wilhelm von Leibniz and Benedict (Baruch) Spinoza. The Cabala of François-Mercure von Helmont inspired Leibniz's theory of monads.

Von Helmont was in close contact with Cabalists such as Henry More and Baron von Rosenroth. In 1671, Leibniz made direct contact with von Helmont. Leibniz was already familiar with von Helmont's work *De alphabeto naturae* and had cited it in his *Nova methodus* in 1667. In 1694, in letters to Elizabeth, Leibniz commented on von Helmont's work *Divine Being*. These letters contained some theological views. Later Leibniz and von Helmont started to correspond.

In von Helmont's work *Opuscula philosophica quibus continentur principia philosophiae antiquissimae et recentissimae* (1690), the term "monad" appeared as originating from the Cabala: "*Divisio rerum nunquam sit in minima mathematica, sed in minima Physica: cumque materia concreta eousque dividitur, ut in monades abeat Physicas, qualis in primo materialitatis suae statu erat, videatur de productione materiae Kabb. denud. [Cabbala Denudata]* (tom. 1, part. 2, p. 310, sqq, et tom. 2 de tract. ult. p. 28, N.4.5) *tunc iterum suscipere apta est suam activitatem, fierique Spiritus: prout in cibis nostris evenit.*" Anne Becco is of the opinion that Leibniz gave his own rational interpretation to the Cabalistic conception of the monad: his monadology arose from the magic of the monad.[40]

Spinoza based his assumed pantheistic vision of God and the cosmos upon the Cabalistic tradition of interpretation of the views of the Ancient Jews. The Cabalists thought that Judaism was originally pantheism.[41]

Besides the theoretical aspect, in which we can see the influence of different schools of Greek and Hellenistic philosophy, the Cabala had a practical side that expressed a desire for domination over the whole world. Supposedly, occult knowledge of letters and numbers would make this possible. The Cabala joins the Reformation and Hermeticism to form the modern paradigm of culture as domination over the world and cosmos.

The syncretism of the age allowed these schools of thought to interact without obstacles. The Cabalistic cult of numbers, especially, encountered the new mathematical picture of the world in which scientists would study the world in terms of numbers, for practical, not theoretical, ends. Paradoxically, during the Enlightenment the prior Renaissance celebration of Judaism would transform into a negative reaction within philosophical and scientific circles.[42]

Twenty-Four

SCIENCE AND UTOPIA: FROM THE HOUSE OF SOLOMON TO THE ROYAL SOCIETY

Utopian literature, inspired more by the *Bible* than by Plato's *Republic*, started to appear at the beginning of the sixteenth century. The term *"Utopia"* was a neologism constructed of the Greek words *"ou"* ("not") and *"topos"* ("place"). So, a utopia was a non-existent place, no place. Other names existed, such as *"Nusquama"* ("Nowhere") or *"Udepotia"* ("Never").[1] St. Thomas More was the author of the first utopia. The exact title of More's utopian work was *De optimo rei publicae statu deque nova insula utopia libellus vere aureus* [*A Truly Golden Short Book about the Best State of the Republic and the New Island of Utopia*].

Initially, utopias were only literary fictions. Over time they took the form of real social programs. Eventually they became specific political programs.

Utopias aimed to illustrate a happy and perfect society. Science also had a special place in the utopian societies. Francis Bacon's *New Atlantis*, the literary counterpart to his philosophical work *Novum organum*, was one of the most influential utopias in this regard. P. M. Rattansi notes that Bacon's utopia bore many similarities to utopias of a clearly Hermetical character, such as the *Reipublicae Christianopolitanae descriptio* [*Description of the Christian Republic*] (1619) of Johann Valentin Andreae, and the *De civitate solis* [*On the City of the Sun*] (1623) of Tomaso Campanella.[2]

In this work Bacon described life on the island of Bensalem, corresponding to the lost Atlantis, where dwelled a happy society isolated from, and unknown to, the rest of the world. The House of Solomon, also called the College of the Six Days of Creation, was one of its institutions.

Bacon called this special place "the very eye of the kingdom."[3] King Solomona (a deliberate distortion of the name of the Biblical King Solomon) founded the House of Solomon 900 years before the visitors accidentally landed on the Island.

The institution was the noblest undertaking in the world. Its purpose was to study God's works and creatures. Because of its antiquity the house of Solomon did not take as models Plato's Academy, Aristotle's Lyceum, or Medieval universities. The House contained fragments of alleged works of the Hebrew King Solomon including a *Natural History* describing all plants.

The task of the House (or College) was to discover the true nature of all things.

King Solomona prohibited all contact between the inhabitants of the island and the outside world, except for special missions that he sent out every 12 years on two ships. On each ship were three brothers who belonged to the House whose task was to learn what was happening in the outside world, and especially to collect knowledge of the latest accomplishments in learning, art, industry, and inventions. They were to bring back books, instruments, and samples.[4] The chief purpose of the House was to learn about causes and secret motions to expand the boundaries of human empire and the ability to influence all things.

As Bacon said, "The End of our Foundation is the knowledge of Causes, and secret motions: and the enlarging of the bounds of Human Empire, to the effecting of all things possible."[5] Science primarily aimed at domination. It had no subordination to a disinterested pursuit of truth.

The House of Solomon had proper equipment for its purpose: (1) special caverns in which they preserved different bodies, produce, new artificial metals, cured diseases, and prolonged life; (2) furnaces for producing porcelain; and (3) high towers for observing meteors and atmospheric phenomena.

The House had fresh- and salt-water lakes for cultivating fishes and birds; dammed streams providing power for different processes; artificial wells; and fountains; great rooms for presenting atmospheric phenomena; an apartment of health that tested the quality of the air, and large therapeutic pools; and different gardens, parks, and orchards.

The gardeners were not concerned with beauty. They were concerned with which soils best suit which plants. They artificially accelerated growth, artificially obtained fruit of greater sweetness in a variety of colors, tastes, and shapes, and created new kinds of plants.

The parks were full of birds, studied and tested in detail. Researchers studied how the birds reacted to different poisons, how to increase their fertility. And they mix species to produce new species.

The House had mechanical devices to make different objects and different kinds of furnaces; and smaller houses of: perspective in which they performed optical experiments; sound in which they studied the qualities and production of sounds; perfumes; mathematics; sensory illusions; and houses equipped with machines.

The personnel in the House of Solomon were divided into groups according to their specialties. Twelve workers, "Merchants of Light," sailed to other lands to acquire books, ideas, and models.

Three workers called "depredators" collected the experiments described in all the books. Three "mystery-men" collected the results in the mechanical arts, liberal arts, and procedures that do not fall under any particular art. Three workers, "pioneers" or "miners," performed new experiments. Three compilers gathered together the results of the previous groups and recorded them in an orderly fashion in tables. Three benefactors or dowry-men studied how to draw benefits from the experiments of the previous groups for the good of humankind. Three workers, "lamps," studied past experiments and directed new experiments to explore more profoundly the nature of things. Another three "inoculators" executed these experiments and reported the results. Finally, three "interpreters of nature" raised the discoveries by these experiments into greater observations, axioms, and aphorisms.

The House had many students, apprentices, and assistants. They held consultations in which they decided which discoveries could make public and which they should keep secret. It had several galleries to present inventions and monuments to those who made discoveries. Representatives from different cities of the kingdom visited the House.[6]

The scientists who worked in the House of Solomon had the status of priests. But the scientific procedure they employed was not like a religious ritual. It was increasingly secular.[7]

Scientific study of nature provided the ability to dominate nature. This increased the power of the island-dwellers. For wealth increases by the development of trade.[8]

After Bacon's death in 1641, Samuel Hartlib, John Comenius (Jan Amos Komenský) and some members of the British Parliament tried to promote the idea of a reform in education, including the foundation of a House of Solomon. The outbreak of war prevented them from realizing this project.

But Bacon's thought was the inspiration for the establishment of the Invisible College and the quite influential Royal Society.[9] The scientific environment of the House of Solomon was international and not under complete State control. It became the model for a freemasonry of collaboration in science, especially among Rosicrucians.

William Eamon maintains:

Solomon's House was not, by any means, a model for a scientific society in a modern liberal democracy, for in Bensalem all scientists 'take an oath of secrecy for the concealing of those [experiments] which we think fit to keep secret; though some of those we do reveal to the state, and some not.' Scientists in Bensalem owed allegiance neither to class nor country but were members of an international scientific freemasonry.

In turn, Pierre Montloin and Jean-Pierre Bayard show correlations between Bacon's utopia and the structure of the activities of the Rosicrucians. They write that the *New Atlantis* is a utopia completely consistent with the mission of the Rosicrucians. This is the ideal of Scottish Rite Masonry. A group of people arrived at Bensalem by a ship that, by chance, was in the vicinity. They were allowed to take part in civil and family life. But some instituions concealed their operations from them. The most important was the Temple of Solomon.

This association made the people happy and concealed nature's secrets. The brothers were secretly dedicated to science. They were divided into classes. Each class had a specific field of work. They gathered together their work and discussed the whole. Three brothers, called the "three lights," dedicated themselves to the experience of the "more sublime light" by which they could profit from their findings.

At their meetings they decided what they could reveal to the broader public, and what they would conceal. Novices and terminators were in the Temple. The

Temple had hymns and a liturgy. Collections, instruments, existed. And, in one house, they could use a machine to show different natural effects that appeared to be marvels.[10]

Characteristic of the work is that the director of the "House of Strangers" wore a white turban "with a small red cross on the top." This sign suggests a connection between the House of Solomon and the Rosicrucians. Francis A. Yates maintains:

> The ideal state or city which Bacon describes was a Christian-Cabalist community. They had the sign of the Cross (a red cross) and the Name of Jesus, but their philosophy was not normal Christian Orthodox philosophy, of any persuasion. It was the occult philosophy, half suspected of being magical, really good-angelical, and more powerful than normal philosophies. Yes, certainly more powerful because it is the Baconian science. The program of learning and research set out in half-mythical, half-mystical, form in the *New Atlantis* is really the Baconian program for the advancement of learning which finds a congenial setting in what one can now recognize as a Christian Cabalist utopia.

Yates adds that the laws that governed society on the island of Bensalem came from Moses and were transmitted with the help of a secret Cabala. In short, the *New Atlantis* was a Christian Cabalistic utopia.[11]

Komenský (*Comenius* in Latin) informally founded the Invisible College. He intended it as an international association of scientists bound together by "holy laws" and the desire for universal wisdom or "*pansophia.*" The outbreak of the Civil War prevented the English Parliament from formally recognizing the *Invisible College.*[12]

The Royal Society formally started in 1660 in England during the reign of King Charles II. Its full name was "*Regalis Societatis Londini pro scientia naturali promovenda*" ("Royal Society of London for Promoting Natural Science"). The Royal Society most often presents itself today as a modern and progressive institution for the study of nature. The statutes state its purpose: "*studia ad rerum naturalium artiumque utilium scientias experimentorum fide ulterius promo-vendas in Dei Creatoris gloriam, et generis humani commodum applicanda sunt*"—they should make efforts to advance the sciences concerning natural things and the useful arts relying on experiments for the glory of God the Creator and the benefit of humankind.[13]

The Royal Society saw itself as promoting the dominance of physical science and a departure from metaphysics, religion, and prejudices. The program of the Royal Society was not as ambitious as that of the House of Solomon. The twentieth century would carry out Bacon's testament be carried out (*New Atlantis* was Bacon's last work and was not finished).[14]

Its founders and members included names recognized to this day as symbols of the scientific attitude that broke out from the "darkness of the Middle Ages." Sir Isaac Newton (1643–1727) was one of the best known. Nonetheless, the Royal Society had a side apparently as backward as the Middle Ages, which they held in contempt in their speeches and writings. K. Theodore Hoppen has written a two-part article in which he has reported some astounding facts.[15] Hoppen shows that, beyond the writings of Bacon, Hermetic and alchemical works influenced the Royal Society. Many of its members held to the Renaissance humanist idea of an Ancient wisdom (*prisca sapientia*). And some even indulged in sorcery.[16]

Hartlib was part of an alchemical circle that included Robert Child, Frederic Clodius, George Starkey, Peter Stahl, and Robert Boyle. Boyle worked with sympathetic medicine and amulets and tried to produce the Hermetic "prime matter" (*prima materia*).[17]

Newton was greatly interested in Boyle's work and called him "a true Hermetic Philosopher."[18] Newton was keenly interested in Hermeticism and his library held many works and extracts of works from authors in that tradition, such as Maier, Paracelsus, Ripley, Basil Valentine, Elias Ashmole, Thomas Vaughan, Robert Fludd, and John Dee.

Newton ascribed to the idea of the Ancient wisdom. This was reflected in his views that he shared with the followers of Hermeticism. He thought that the Ancient Egyptians knew the Copernican system, that Ancient peoples knew of the atomic structure of matter, and that the Phoenicians knew of the effects of gravity on the movement of matter. More than half of Newton's works were about alchemy and religion, not about science.

Hence Richard H. Popkin tells us: "The historians of science can read Newton and worry about why he introduced metaphysical issues, and why more than half of what he actually wrote is about strictly religious matters and alchemical matters, including the interpretation of *Daniel* and *Revelation*, but scientists and science students can stay in a twentieth-century context with twentieth-century ideas."[19]

Still, we can draw no sharp line between what some people regard today as Newton's purely scientific ideas and his other interests. Hoppen maintains that Newton's intense study of alchemy largely resulted from "his belief in the existence of an Ancient wisdom, which had been lost and which he was in the process of rediscovering."

Newton believed that the Ancient Egyptians had known and taught Copernican astronomy and were familiar with matter's atomic structure and its gravitational motion through a spatial void. He traced their knowledge of these matters to Moschus the Phoenician. Likely, Newton considered alchemy as part of the Ancient wisdom. He thought that the mysteries of this Ancient wisdom was open only to the initiate.

Hoppen claims that this approach "informed much of Newton's philosophical research, and it is wrong either to deny the reality of his esoteric interests or to separate them entirely from his 'official' scientific studies."[20]
Hence, while some contemporary thinkers present Newtonian physics in the conceptual categories of the mathematics and physics of our time these categories do not show Newton's entire picture.[21]
John Locke's library held more than 60 works on alchemy. Locke was in close contact with Hartlib's circle. Newton's Cambridge University colleague, Henry More, regarded Cartesianism and Platonism as fragments of the Jewish Cabala. And his own system was a combination of Cabalistic commentaries on the *Old Testament*, Pythagorean mysticism, Platonism, and neo-Platonism. He also accepted occultism, spiritualism, and sorcery.[22]
The Royal Society granted membership to Gottfried Wilhelm von Leibniz in 1673. Popkin tells us that Leibniz was interested in alchemy and maintained contact with several alchemists, including Oldenburg, Athanasius Kircher, and Michael Sendivogius:

Leibniz, who was elected F. R. S. in 1673 and who corresponded with Oldenburg, also corresponded with Athanasius Kircher, and was a student of the writings of the German alchemist, J. J. Becher, whose process for "making iron" from non-metallic substances, he thought "must be accepted." In 1671 he wrote to Oldenburg urgently requesting a copy of "a certain book, in English, by Gabriel Plattes, dealing with subterranean matters" in which there was "described some process or other for extracting a salt from antimony which . . . makes the truth of transmutation quite clear." In the same letter Leibniz shows an acquaintance with the writings of the Polish alchemist, Michael Sendivogius.[23]

Leibniz's alchemical interests predated his membership in the Society. At twenty years of age in 1666, Leibniz had become the secretary of an alchemical association in Nuremberg that was connected with the Rosicrucians. About this connection, Karl R. H. Frick states:

The philosopher Gottfried Wilhelm von Leibniz (1646–1716) temporarily belonged to such a society. At twenty years of age in 1666 he went to Nuremberg, where his maternal uncle, the clergyman Justus Jacobus Leibniz, brought him into an alchemical society that existed from 1654 to 1700. To all appearances it concerned a group that belonged to the Order of the "Inseparable Ones." As their secretary, the young Leibniz made excerpts from the alchemical writers who had to make minutes of the works they did in the Society's laboratory and had to do correspondence. He exercised this function, however, only for a year. By his activity he appeared to discover no "true adept" or a "genuine Rosicrucian," when he wrote in a later letter of March 26, 1696: "I suspect that the brothers of the Rosy Cross were fictitious, which Helmontius confirmed for me. For to know what things are happening in remote places, how to make oneself invisible and invulnerable, without doubt are jestful or ridiculous things.

(*So gehörte u. a. auch der Philosoph Gottfried Wilhelm von Leibniz* (1646– 1716) *vorübergehend einer solchen Gesellschaft an. Mit 20 Jahren kam er 1666 nach Nürnberg, wo ihm sein Oheim, der Pfarrer Justus Jacobus Leibniz, in eine alchemistische Gesellschaft einführte, die dort von 1654 bis 1700 existierte. Allem Anschein nach hat es sich um eine Gruppe gehandelt, die dem Orden der 'Unzertrennlichen' angehorte. Als ihr Sekräter hat der junge Leibniz alchemistische Schriftsteller exzerpiert, die im Laboratorium der Gesellschaft vorgenommenen Arbeiten protokollieren und die Korrespondenz führen müssen. Diese Funktion übte er allerdings nur ein Jahr aus. Bei seiner Tätigkeit scheint er weder 'wahre Adepten' noch 'echten Rosenkreuzer' entdeckt zu haben, wenn er in einem späteren Brief vom 26. März 1696 schriebt:* 'Fratres roseae crucis fictitios fuisse suspicor, quod Helmontius mihi confirmavit. Nam scire, quae remotis locis fiunt, invisibilem sese atque invulnerabilem reddere, haud dubie nugacica vel potius irrisoria sunt.'")[24]

The scientists of the Royal Society active in Hermeticism and alchemy included Thomas Henshaw, Thomas Vaughan, William Oughtred, Sir Robert Moray, Athanasius Kircher, John Winthrop, Sir Kenelm Digby, Nicholas Le Fèvre, Robert Plot, Edmund Dickenson, Theodore Kerckring, and Christopher Baldwin. Many others existed.

These authors looked to Ancient Egypt and the Mosaic tradition in their search for a new way of looking at nature. Frequent links existed between the members of the Royal Society and the Rosicrucians.

For example, Hoppen says that Dickinson published a work called *Physica vetus et vetera: sive tractatus de naturali veritate hexaametri Mosaici* [*Old Physics and Old Things: Or a Treatise on the Natural Truth of the Mosaic Hexameter*] (London: 1702). In this work, Dickinson said that Moses was the father of the philosophy of nature because he knew how to dissolve the golden calf by chemical compounds and possessed all the wisdom known to the Ancient Egyptians. In turn, many of Moray's contemporaries recognized him as one of their greatest chemists, an excellent mathematician, and a great patron of the Rosicrucians.[25]

Some of the primary models for the modern institutes of science and research reflected non-European and non-Christian esoteric traditions. Some of the Royal Society's most vigorous minds appear to have consciously worked in an esoterically influenced, alchemical-like laboratory to recast Western culture into a new substance.

The new conception of science played an essential role in this process and became an instrument for changing Western civilization. Apparently, retouched views represented by symbolic names were to provide credibility to society for the direction of the changes.

Twenty-Five

A NEW MODEL OF SCIENTIFIC KNOWLEDGE

1. Science without Causes?

The search for causes was the chief motive in Ancient Greek scientific thought. This pursuit reached its apex in Aristotle's philosophy. Book 1 of Aristotle's *Metaphysics* analyzed the views of his predecessors. He showed that, when we seek a scientific answer to the question "*dia ti?*" ("because of what?"), showing a thing's material or formal cause is insufficient. Scientifically to understand reality we must know all the proper causes of a thing. Beyond the material and formal causes we must also know the efficient and final causes.

Greek science was etiological (from *aitia*—cause). The Hellenic and Medieval (Jewish, Muslim, and Christian) ages borrowed this conception from the Greeks. At the end of the Middle Ages nominalism made a strong assault against the etiological conception of science.

Aristotle thought of the efficient cause and final cause as different aspects of the formal cause. As Joseph Owens has said, "[I]n some sense every type of causality finds its ultimate explanation in the form." An efficient cause is a form as it acts on another being (or upon itself as other). And the final cause is the form as a perfection where the end or term is to arrive at the fullness of a nature. The material cause, in turn, is a potentiality completely disposed toward a form.[1] Without form, the Greek conception of being and science falls apart.

Nominalism primarily attacked the idea of form. Necessarily, this attack led to the fall of metaphysics and the Greek conception of science.

Francis Bacon's writings provide a clear example of how modern Western physical science abandoned etiology. Bacon writes in his *Novum organum*: "It is a correct position that 'true knowledge is knowledge by causes.' And causes again are not improperly distributed into four kinds: the material, the formal, the efficient, and the final."[2]

This formulation is most orthodox, but in the following sentence we see that it was merely a rhetorical device: "But of these, the final cause rather corrupts than advances the sciences, except such as have to do with human action. The discovery of the formal is despaired of. The efficient and the material (as they are investigated and received, that is, as remote causes, without reference to the latent process leading to the form) are but slight and superficial, and contribute little, if anything, to true and active science."[3] Bacon rejected all the causes: final, formal, efficient, and material.

Before we present the alternative conception of science diametrically opposed to the Greek conception, we should consider Bacon's criticisms. He linked the concept of final cause to the mode of human action. When we human beings have an end in mind, we seek the proper means to achieve his end. Bacon maintains that we make a mistake to extrapolate this conception of teleology to non-human activity. And Bacon was right to reject this approach.

However, the metaphysical conception of teleology is different: an end is a motive that determines action. In metaphysics the questions of reflection on the end as in human activity and the achievement of the end (the end as terminus) are secondary. Metaphysics focuses on the end as the motive that guides action. Without the end in this sense, we cannot provide a metaphysical explanation for the fact of change.

For activity to arise in place of inactivity (in every activity a phase of the beginning of activity occurs where previous inactivity had existed), a reason of being for the activity, and for its direction or determination, must exist. The end, understood as the motive of activity, is the reason of being for the beginning of activity, just as exemplary forms (or a nature that exemplary forms determine) are the reason of being for the direction or determination of activity.[4]

Bacon worked with an anthropomorphic conception of the end and was able easily to reject teleology in this guise and eliminate it from science.

Bacon said that to search for forms is hopeless because forms are simply "figments of the human mind."[5] Hard to find something in a thing that is not there. And someone who persists in such a search engages in a hopeless task. Bacon presented no metaphysical critique of the concept of form.

He referred simply to the nominalistic tradition, and specifically to conceptualism. Form as "a figment of the human mind" is a concept without any counterpart in things. Bacon definitively states that the human mind makes a mistake to treat form as the most important thing.[6]

Bacon rejects the efficient and material cause as vague and superficial because they are fluid. The same property occurring in a body may have different causes.[7]

The concept of "cause" that Bacon employs here is quite narrow and has no bearing on the metaphysical conception of efficient or material cause. Bacon works with a fictitious version of causality far removed from its essential metaphysical sense.

While some people regard Bacon as a leading critic of Aristotelianism, he did not know Aristotle very well. At the age of twelve, Bacon started studies at Cambridge University. He interrupted his studies two years later (April 1573–December 1575), for eight months, because of the plague.

His course of studies was thus quite short. And the spirit of nominalism pervaded the University.

If we reject causes what will be the basis of scientific knowledge? Bacon shows a new scientific way based upon two pillars.

(1) Instead of form we have law. Instead of searching for fictitious forms science should discover laws. Bacon writes: "For though in nature nothing really exists besides individual bodies, performing pure individual acts according to a fixed law, yet in philosophy this very law, and the investigation, discovery, and

explanation of it, is the foundation as well of knowledge as of operation."[8]
The nominalistic approach to reality found a new foundation for science. Law, not form, became science's objective point of reference in knowledge. The scientific question changed: instead of "why" (*"dia ti?"*), the new question was "how," in the sense of a description to answer this question the scientist presents the law that the described changes follow.

(2) In the new conception of science. Bacon writes, "[I]nquiries into nature have the best result when they begin with physics and end in mathematics." Herbert Butterfield tells us:

We learn that Kepler's discovery of the laws of planetary motion was made possible only by the fact that he inherited and developed further for himself the study of conic sections, a study in which he excelled all of his contemporaries. . . . At a later date the same phenomenon recurs and we learn that the problem of gravitation would never have been achieved without, first, the analytical geometry of René Descartes and, secondly, the infinitesimal calculus of Newton and Leibniz. Not only did the science of mathematics make a remarkable development in the seventeenth century, but in dynamics and in physics the sciences give the impression that they were pressing upon the frontiers of mathematics all the time. Without the achievements of the mathematicians the scientific revolution, as we know it, would have been impossible.[9]

Quantity comes to center stage beside law. To know something scientifically is to discover laws that we express mathematically. Mathematically-grounded physics supplants metaphysically-grounded physics. Science searches for quantitative laws. We cannot separate the question of the descriptive "how" from the question "how many." Law supplants form, quantity supplants matter, while we eliminate final and efficient causes.

Inclusion of mathematics in the structure of scientific knowledge leads to a modified concept of experience. Scientific experience becomes scientific because we can express such experience mathematically. For this to take place, we must sufficiently develop the proper branches of sciences. The conception of science was still maturing during the mid-seventeenth century because René Descartes' analytic geometry and Gottfried Wilhelm von Leibniz's infinitesimal calculus were necessary. Only after experience became wedded to mathematics could we speak of the scientific revolution of a basic change in the conception of science.[10] Experiment or mathematics alone does not constitute science. Science consists in the experimental method wherein we can translate experiments into the language of science: mathematics.[11]

Science uses experiments in its search for laws. The physical scientist must express experiments and the discovered laws in mathematical language. The scientist needs the proper instruments to measure his experiments. And the new conception of science brings with it the development of new instruments. The scientist does not borrow these instruments from other fields. The scientist builds them specifically according to the demands of science and intends them to serve science. The experimental method requires instruments of measurement.

When we conceive of science in this way and we properly equip science, science allows us to read the Book of Nature which, as Galileo Galilei said, is written in the language of mathematics.

2. Induction: From First Premises to the Syllogism

Francis Bacon was fully aware that he was undermining the scientific value of traditional induction. He introduced a new kind of induction in its place. Hence, he wrote:

> But the greatest change I introduce is in the form itself of induction and the judgment made thereby. For the induction of which the logicians speak, which proceeds by simple enumeration, is a puerile thing, concludes at hazard, is always liable to be upset by a contradictory instance (*instantia contradictoria*), takes into account only what is known and ordinary, and leads to no result. Now what the sciences stand in need of is a form of induction which shall analyze experience and take it to pieces, and by a due process of exclusion and rejection lead to an inevitable conclusion.[12]

Enumerative induction, a simple act of counting, is quite a meager type of induction. And its main fault is that it does not protect us from a contradictory instance.

Induction enables us to form concepts that become elements of propositions. These, in turn, become elements of a syllogism. Faulty induction weakens the value of the syllogism as a whole.[13]

Bacon's move was a strike at the weakest link in Aristotelian syllogistics. A syllogism should ultimately refer to premises that are true in every case, which refer to a thing's essential nature of things, and are universal. Where do these premises originate, and how can we be sure they meet these requirements?[14]

Too hasty generalization causes our thought to be infected with different errors, which Bacon calls "idols."[15] Bacon thought that traditional induction had serious shortcomings. So, he proposed a new type of induction:

> [T]he induction which is to be available for the discovery and demonstration of sciences and arts, must analyze nature by proper rejections and exclusions; and then, after a sufficient number of negatives, come to a conclusion on the affirmative instances—which has not yet been done or even attempted, save only by Plato, who does indeed employ this form of induction to a certain extent for the purpose of discussing definitions and ideas. But in order to furnish this induction or demonstration well and duly for its work, very many things are to be provided which no mortal has yet

thought of; insomuch that greater labor will have to be spent in it than has hitherto been spent on the syllogism. And this induction must be used not only to discover axioms, but also in the formation of notions. And it is in this induction that our chief hope lies.[16]

Thus we will base scientifically valid induction upon a consideration a sufficiently great number of instances where we can properly rule out negative cases and derive conclusions from positive instances. Bacon formulated canons for so-called "eliminative induction." Whenever Plato established a definition, he presented contrary instances until the moment when he could bring no contrary against the proposed definition. But in Bacon's case we find a more precisely formulated relation between different possibilities as mutually exclusive. By elimination, one from among the different possibilities remains because of the relation of disjunction among the possibilities.[17]

Bacon applied induction as a method of scientific knowledge and presented a theory of induction. His analysis of the Aristotelian conception of the syllogism noted that the middle term is the heart of the syllogism. But we do not know the middle term syllogistically.

Is discovery of the middle term scientific? Does it occur by reason's natural talents, or is its discovery a question of luck? If the second, the syllogism loses its scientific value.[18] The traditional syllogism does not meet the expectations of Baconian science. It fails to provide a proper method for establishing primary premises. It does not lead to conclusions that would allow us to make discoveries. Given the insufficiency of building a syllogism correctly from a Baconian formal point of view, and given the conclusion that the method upon which we base the crucial first premises should be scientific, Bacon concluded that Aristotelian induction is not scientific.

At the moment when Bacon developed his scientific procedure for establishing first premises, he encroached upon the territory of metaphysics and epistemology. His critique of Aristotelian induction became a critique of Aristotelian metaphysics. In this way when Bacon discussed induction and the syllogism, he was starting to do metaphysics; and from an empirical and nominalistic position. Bacon's theory of induction ultimately arose from a non-Aristotelian metaphysics. And this is the main reason for the disagreement between Bacon and Aristotle.

No essential disagreement exists between Bacon and Aristotle regarding the structure and purpose of induction. We must only distinguish between spontaneous induction and scientific induction. Both authors speak of the role of experience at the starting point, and both warn against premature generalizations. Bacon wrote in greater detail of the danger of premature generalizations. But, in this matter, he did not disagree essentially with Aristotle.

Mieczysław Albert Krąpiec maintains that Bacon's claim that Aristotle "did not know the instrument of inductive thought, and that his only method was syllogistic deduction" is false. Krąpiec cites in Aristotle's defense Edward

Zeller's observation "that Aristotle was one of the most experimental philosophers." One of Aristotle's basic principles was that we should gather as many facts as possible." For this reason, Zeller maintains, Aristotle's writings contain "much empirical material."

On the basis of Aristotle's and St. Thomas' texts Krąpiec goes so far as to present three stages in thought-formation based upon inductive sciences: (1) Since the basis for these sciences is always experience broadly considered, we start by "gathering as many observed facts as possible." As we do this, "we must refrain from premature generalization and premature systematization." (2) We must subsequently filter the empirical data to discover our general principle and remove "additional elements loosely or accidentally connected with a general principle we have discovered." (3) "We must then investigate the stable elements connected with a thing's nature." According to Krąpiec, Bacon and John Stuart Mill developed this stage of induction more extensively than did Aristotle and St. Thomas.[19]

3. Experience and the Discovery of Nature's Secrets

Mysteries exist in society, secrets that institutions such as the State or Church guard closely. These institutions have laws to prevent unauthorized persons from intruding into the private matters of others. Traditionally human beings regarded nature as a subject that possessed the right to keep her own secrets. During the Renaissance a brutal invasion ensued into the secrets of the State, Church, and nature. In *The Prince* Niccoló Machiavelli intended to reveal the mysteries of ruling the State; Martin Luther intended to lay bare the mysteries of the Church; and Bacon proposed to lay bare Nature's mysteries.[20]

The call to extract and make public mysteries previously protected by laws and institutions would lead to changes in society and civilization in general. These changes affected science. And Nature was subjected to procedures intended to extract her mysteries. These procedures took the form of experiments.

Initially the term "experiment" was a word used to refer to physical torture. If Nature as an object was to meet the demands of the new science, contemplation and discussion of Nature were not enough. Bacon thought that the scientist needed to subject nature to torments in the workshops of craftsmen to make her give up her secrets.

Charles Webster maintains "Bacon had different reasons for preferring scientific investigations to concentrate on subjects already treated by the craftsman." In Bacon's time, Webster claims, mechanical knowledge of materials and techniques was already advanced, especially in areas like metallurgy and horticulture. This rudimentary mechanical knowledge provided a solid foundation for further experimental investigations. Nature would reveal her full potentialities when "forced out of her natural state, and squeezed and moulded" in craftsmen's workshops. Thus, in the modern age, Nature became like Proteus, who was induced to reveal his true shape only when straightened and held fast. Similarly it was necessary to submit nature to the trials and vexations of art. As a lawyer Bacon naturally applied familiar professional terminology to the activities of the craftsman. By "interrogation" applied with extreme determination and cunning, nature would be "tortured" into revealing her secrets; she would then submit to voluntary "subjugation."[21]

Nature, in short, became viewed as a slave. According to the tradition of Roman law, the testimony of a tortured slave is valid. The mechanistic picture of the new scientific world reduced mercy or pity. Cutting up a living animal would mean no more than taking apart a clock, since both are merely machines.

At that moment, in the name of scientific progress, improvement of human life, the process of the violent degradation of the natural environment started on an unprecedented scale. Scientists started to extract Nature's mysteries by torturing Nature.

Today, we talk much about ecology. If we want to see a real reversal in the devastation of the physical world, if we will not settle for a propaganda victory, we must address the chief sources that are poisoning our civilization: some ideas. We must consider how we understand Nature. Is Nature simply a mechanism, and can we learn the laws of nature only by tormenting Nature?

4. Discovery and Marvels

Paracelsus was a herald of the idea of progress in science. He thought that the Ancient peoples and the Arabs had still not discovered everything and that science could still make new discoveries. He assumed that God desires that we should allow nothing to remain concealed, and that human beings should complete what God had left unfinished. Science discovers, and alchemy completes. Paracelsus thought that the epoch of the Holy Spirit would soon arrive in which, according to Joachim of Fiore, God would reveal and make known everything, and would bring the arts to perfection.[22]

In the contemplative approach to the object of scientific knowledge, an investigator looks for the general, unchanging, and necessary. In the utilitarian approach these three features change. Instead of proceeding from what is individual to what is general, the new approach focuses on the concrete thing insofar as it is concrete in the hope of finding "marvels" (*mirabilia*). These marvels arise from the action of hidden forces, sympathies, and antipathies. Knowledge of these forces provides a great advantage to the initiate since he or she can bring about similar effects.[23]

Twenty-Six

PHILOSOPHY AND GNOSTICISM

Gnosticism is a mysterious religious and intellectual movement that no one has yet adequately studied. It started to take shape in the second-century A.D. in the Hellenistic world. Gnosticism was syncretistic and even parasitic. Many varieties of it existed, and to regard it as a single uniform teaching is difficult. Unlike the great religions, Gnosticism has no clearly identified founder who proclaimed a clear set of doctrines.[1]

Gnosticism contains many myths and religious currents: (1) mythological and religious elements from Babylon, Egypt, Judaism, and Christianity; concepts drawn from Greek philosophy. Gnosticism is no religion or philosophy. But it has competed with religion and philosophy. As a movement, Gnosticism was able to penetrate Western culture, philosophy, and religion, especially Christianity.

Early Christian apologists saw the dangers of Gnosticism. St. Irenaeus of Lyons compared it to Greek mythology's Hydra: when we unmask one falsehood, hundreds more appear. Western culture still suffers from Gnosticism's infection. Gnosticism has been a major influence in art, politics, psychology, and some schools of modern and contemporary philosophy.

Some philosophers have rejected Aristotle, Scholasticism, and Christianity, only to fall into Gnosticism. Sometimes Gnosticism has only an indirect influence on philosophers. And Gnostic elements in a philosopher's thought do not always appear explicitly in his published writings. Georg Wilhelm Friedrich Hegel (1770–1831) and Friedrich Schelling (1775–1854) are examples of philosophers who deliberately concealed the Gnostic sources that inspired them.

Gnosticism is no philosophy that we can identify with a founder, as can Pythagoreanism, Platonism, and Aristotelianism. Gnosticism is a set of ideas and a method that appears sporadically in Western thought, where it takes a Western form. But this is only an external guise. For this reason, Irenaeus compared Gnosticism to a Hydra, and some writers speak of its different metamorphoses.[2] Gnosticism can assume flesh in different teachings and systems.

Gnosticism's infection of modern and contemporary culture is all the more curious when we consider how a minimalist approach is increasingly common in contemporary physical science. Physical science tends to shun rationalism for sensualism, and to reduce experience to sense impressions. The prevalent attitude in physical science is to oppose metaphysics and pure speculation. The prevalent minimalism in physical science actually makes Gnosticism easier to infiltrate philosophy by giving it an air of mystery and novelty. Gnosticism is an intellectual trap. And an elite who mistake an intellectual labyrinth for truth are especially susceptible to falling into it.

1. What is Gnosticism?

The study of Gnosticism is difficult because of its inherent complexity and because, for a long time, all we knew about it came from Christian apologists, its adversaries. Christians destroyed most Gnostic writings, which threatened Christianity. Finally, in 1945, investigators discovered some Gnostic books in Nag Hammadi, Egypt. Publication of these books occurred thirty years later. These books partially filled the great gap in our knowledge.[3] The books confirmed that the Christian apologists who fought Gnosticism had faithfully presented its teachings. The main lines of their interpretation of Gnostic doctrine matched what was in these books.

The Greek term "gnosis" means "knowledge," and a "*gnostikos*" is a person who possesses knowledge. Plato uses the term "*gnostikos*" when he defines theoretical knowledge as distinct from practical knowledge and skills.[4] During its inception in Greek culture, the Greeks identified *gnosis* with scientific knowledge and the crystallization of *theoria*.

The kind of *gnosis* that appears in the second-century A.D. is of a completely different character. It arose when civilizations collided, and it contained mythological, religious, poetic, and moral elements. These elements were so intertwined that, to this day, scholars cannot agree on a single definition of Gnosticism. Different scholars emphasize different aspects that they regard as central to this complicated phenomenon.

Gnosis in the Gnostic sense means a specific kind of knowledge that is supposed to bring liberation and salvation. It supposedly comes from a deity who passes it on to a group of chosen individuals. *Gnosis* contains a vision of reality, God, the origin of the Heavens, human beings, and salvation.

This vision is common to all varieties of Gnosticism. We cannot consider it to be a syncretic mixture, although it has many elements that refer to other religions and to Greek philosophy. Syncretism and the presence of mythological elements in Gnosticism have been a major obstacle to reconstructing the essence of the Gnostic teaching and have given rise to different interpretations and versions.

The Gnostic picture of the universe includes the Earth, where humans dwell, seven celestial spheres, and different gods. The Greeks thought of the universe as a place of order, and so described it as a "*cosmos*."

The Gnostics strongly opposed this picture of cosmic order. Their position marked an enormous turning-point in a civilization dominated by Hellenistic thought and had far-reaching consequences.

The Gnostics held that a good God could not cause an evil world. So its author must be an evil god. Some Gnostics thought that this world came from the angels of the last Heaven whose king is Yahweh, the Jewish God.[5] Other Gnostics called the evil god Jaldabaoth. This god thought that there was no higher god above him and acted in anger and pride. The world is evil because of him. For, as the Demiurge, he made an imperfect work.

Different versions of the Jaldabaoth myth have existed. Giles Quispel recounts one version:

The evil creator of the world, Jaldabaoth, inherited from his mother, who was named Sophia, a divine power of creation which he jealously kept to himself. The demon had to be deprived of his divine power and return this power to the kingdom of light. Jaldabaoth was ordered to breathe the divine spirit he had inherited on to the face of the new creature . . . man, Adam. Unfortunately, he later sent angels to fertilize offspring with the dwellers of earth. They grafted on to all people an *antimimon pneuma*, an anti-spirit—as a counterbalance, as it were, for the divine spirit of life—which by the power of pleasure attached people to that which was of earth.[6]

Gnosticism diametrically opposes Greek philosophical thought, orthodox Judaism, and Christianity. The *Old Testament* affirms the goodness of the created world: "And God saw all the things he had made and they were very good."[7] Gnosticism also tells of a war among the gods. The Gnostic Saturninus taught that Christ the Savior came to defeat the God of the *Old Testament*.[8]

According to the Gnostics, the world came from the gods of a lower celestial sphere. It was not created *ex nihilo*. Lower gods produced it from pre-existing matter. This position resembles Plato's teaching in the *Timaeus*, except that Yahweh or some other god takes the place of the Demiurge.

The Gnostics did not believe in resurrection of the body. They taught metempsychosis, transmigration of souls. They opposed the idea of a cosmos, and were anti-somatic, regarding the physical body as the source of evil. Orphic thought, which influenced the Pythagoreans and Plato, regarded the body as a prison for the soul (*soma-sema*, body-prison). But the Gnostics went further than the Orphics and completely condemned the body.

Gnostic mythology presents the human being as a fallen god bound to a body because the soul lusted for pleasures and fell through seven Heavenly spheres to the Earth.

The body and the world are, therefore, evil, as is the god, or gods, from whom the world came. According to the Gnostics, we confront absurdity. Gnosticism implies a radically pessimistic worldview.

2. The Way of Salvation

Gnostics maintain that the only way to salvation is to acquire secret knowledge (*gnosis*). A divinely chosen person connected with the nameless God supposedly possesses this knowledge. The circle of initiates who hear the voice of a chosen God can look forward to deliverance from this absurd world. The secret knowledge brings the elect memory of the first origins of the soul and the awareness of their own divinity.[9] Gnosticism was and is elitist because *Nous*

(God), the intellect with the capacity for intuition, has equally distributed intellect to everyone. So, only some people can share in divinity.[10]

Simon Magus was the first known example of a Gnostic elect. The *Acts of the Apostles* (8:9–12) mention him. Simon used his arts to astound the people of Samaria, who thought he had divine power. Philip converted and baptized Simon. But he wanted to buy the gift of the Holy Spirit. So the early Church condemned and excommunicated him from the community.

Simon's companion was a prostitute named Helena. In Gnostic mythology Simon became a god and Helena became a goddess, the first divine thought (*ennoia*) and the mother of everything whose mind held the plan for the creation of the angels and archangels. Helena was also Wisdom (*Sophia*) and the Soul of the World.[11] Gnosticism has elements of the doctrine of emanation: the celestial spheres emanate in succession from the nameless God, and other spheres emanate from these. Gnosticism also has pantheistic elements: the elect are a part of the divinity.

At a conference in Messina in 1966 scholars agreed that Gnosticism was a doctrine concerning divine mysteries intended for a group of chosen individuals. Gnosticism was a group of teachings in the second century A.D. that held that a human being is a divine spark that came from Heaven and should return to God to become one with him. In Gnosticism, the self, the divine spark, the reality of God, and knowledge of God are identical.[12] Gnosis is the only way of salvation from the world's chaos and absurdity.

During its inception Gnosticism posed a serious threat to Christianity, still a young religion building up its theoretical identity, and beset with different heresies. Theological insight and vigilance of the great apologists, such as Irenaeus, exposed the dangers of Gnosticism. And Plotinus criticized it in philosophical terms. But it never disappeared completely from Western culture. It found a home in Medieval sectarian movements, such as the Cathars, Albigensians, and Bogomili. In modern times Gnosticism re-emerged to make inroads into religion, philosophy, and other parts of culture, such as politics and art.

3. Gnosticism and German Idealism

We can recognize the esoteric subterranean, Gnostic, layer of modern culture by focusing upon the change from the conception of science as *scire propter ipsum scire* (knowing for the sake of knowing) to *scire propter uti* (knowing for production). Many authors and champions of modern science who claimed they were escaping Medieval darkness were also keenly interested in Hermeticism, the Cabala, and alchemy. Esotericism also gained ground in philosophy, even while philosophers claimed to represent pure empiricism or pure rationalism. The Cabala influenced philosophers such as Gottfried Wilhelm von Leibniz and Benedict (Baruch) Spinoza. And German idealism as a whole is the clearest example of philosophers who consciously surrendered to esotericism. We still feel the repercussions today.

German idealism draws upon a line of thought that started with Meister Eckhart and Jan Tauler continued. It also draws upon a version of Gnosticism reframed in rational terms and inspired by the works of Jacob Boehme and other theosophists and Cabalists such as Emanuel Swedenborg and Friedrich Christophe Oetinger.[13]

This rationalization of esoteric doctrine meant that it emerged from its underground hiding-place and gained public recognition in universities. Then it set the tone for Western philosophy and appeared in different philosophical schools. It appeared in many guises, from mysticism to political programs.

Franz von Baader was a professor at the University of Munich. He collaborated with Hegel, and published the works of Eckhart and Boehme. Baader intended to form a synthesis of Eckhart and Boehme and to supplement their thought with the Jewish thought of the Cabala. Baader thought that he could, thereby, provide a remedy for the crisis in philosophy at that time.[14]

This approach attracted Hegel. Baader recalled a conversation he had with Hegel:

> I was often in Berlin in the company of Hegel. One day in 1824 I read to him a text of Meister Eckhart of whom until then Hegel knew only his name. He was so enthusiastic that the next day he gave me an entire lecture of Meister Eckhart and finished with the words: "*Da haben wir es ja, was wir wollen*"—"This is just what we want, this is the whole of our ideas, our intentions." I say to you—Hegel directed his words to Baader—Meister Eckhart was rightly called Meister. He was greater than all other mystics. . . . I thank God for allowing me to learn of him in the middle of the philosophical problems of our epoch. The cry of monkeys against mysticism, so arrogant and stupid a cry, could not irritate me more, and this allows me to embark on the study of Jacob Boehme.[15]

The comparison between Eckhart and Boehme strikes me. Eckhart was a Dominican. The Inquisition in Cologne condemned his works. So did the Roman Inquisition (1328). Many scholars often portray Boehme as a poor tailor and a self-taught man who developed a Hermetic philosophical system.

As Eckhart developed his philosophical views, he arrived at mysticism. This led him to pantheism. He identified the human self with the divine self, and then included the human self in the process of the creation of being. While contemporary Medievalists are not unanimous in seeing Eckhart as a pantheist my problem with Eckhart is not whether or not he was a pantheist. It is that subsequent thinkers, especially German idealists, later interpreted his writings in pantheism's spirit.[16]

Eckhart also developed a basic philosophical terminology in the German language. We can see traces of his terminology throughout the writings of

German mystics. These terms include: *sein, wesen, wesenheit, das seiende, das nicht, nichtigkeit, nichtigen, gestalt, anschauung, erkenntnis, erkennen, vernunft, vernünftigkeit, verstand, verständnis, verständigkeit, bild, abbild, bildhaftigkeit, entbilden, grund, ungrund, ergründen, Ich, ichkeit, nicht-Ich, entichen,* and *entichung*.[17] No surprise that German philosophy, almost by its nature, is heavily mystical and idealistic. Such was the language it inherited. Speculations on the meaning of terms, which they endowed with majestic meanings supposedly corresponding to some language of divine revelation such as Hebrew or Egyptian, necessarily inclines German philosophy far from realism.

Boehme was no poor or spontaneous genius. The portrait of Boehme as a simple tailor has persisted for 300 years on the basis of a carefully nurtured legend. Boehme came from a powerful farming family, never faced any financial difficulties, and was able to devote his adult life to writing. While he had no formal philosophical education, he knew the current literature in alchemy, the Cabala, mysticism, and religion. He traveled extensively, had a circle of learned friends, and was a devout Lutheran.

The system Boehme created was no philosophy in the strict sense. It was more a Gnostic system formulated in the language of Hermeticism and deeply rooted in then-current trends, such as Medieval and Renaissance mysticism, Lutheranism, and the views of such representatives of the Reformation as Caspar Schwenkfeld and Valentin Weigel.

Except for one, Boehme wrote all his works within a period of six years. His views changed in some respect. But an element of Gnosticism was constant and omnipresent.

Boehme presented himself as someone with access to God's essence. He claimed that, through God, he saw all creation. This attitude is typical of a Gnostic master. Boehme recorded his vision in writing and passed it on to adepts for belief. The vision includes views on God, the world, human beings, and evil. As is characteristic of Gnosticism, Boehme treated the problem of evil as the most crucial.[18]

Boehme's first work was *Aurora* [*Morgenröte im Aufgang*]. In it he maintained that things consist of six contrary qualities: hot-cold, bitter-sweet, and bitter-salty. These qualities are dynamic, not static. Each quality has a modality of good and evil. In this way, three parallel oppositions ground reality's structure, and one opposition cuts across and divides reality in two (the dualism of good and evil).

Boehme introduced the dogma of the Holy Spirit in his speculations. God relates to the world as the human soul relates to the body. The Holy Spirit permeates nature, God's body, and directs the modality of good in things. God the Father is the source of all qualities (Heaven). God the Son is the source of eternal joy (the Sun). The Holy Spirit is the principle of motion and life (fire, air, water). The basic qualities of nature exist in God. These basic qualities are seven, corresponding to the number of astral bodies and the base metals in alchemy. The seventh quality results from the other six as they act together. These qualities gain the status of spirits that are sources (*quell-geister*). The world comes from God, who is corporeal and with his corporeality filled the world before creation. Boehme's system oscillated between pantheism and panentheism. He denied creation *ex nihilo* and God's transcendence of the world.

Lucifer is one of three archangels, the others being Michael and Uriel. Lucifer wanted to rival God and fell, pulling with him angels from his sphere. His fall was the cause of the evil in the world. God was not responsible for evil because Lucifer was free. The visible world was God's second creation. He created the world in an attempt to correct the evil that had infected the angelic world. No evil exists in God. But God's structure is such that evils can be actual in the creatures that reflect God, since Lucifer had acted freely.

In a work called *Three Principles*, Boehme tried to explain God's nature as the first cause. In this work he said that God is the undifferentiated source, and His major strength is will. Because God had no other object apart from Himself, he wanted to be divided from Himself. Two centers, light and darkness, resulted. Light penetrated the darkness, and seven rays appeared. Boehme's scheme contains some characteristic elements: God is will; God needs to manifest Himself and this leads to differentiations; evil is an ontological principle.

Boehme distinguished three worlds: (1) Hell, a first principle in God, (2) Heaven, and (3) Earth.

In *Man's Threefold Life*, Boehme described God as the absolute person and recognized God's essence as self-revelation. The appearance of contraries is an effect of God's self-revelation. As pure will God acquires self-awareness by the process of self-revelation.

The tendency contained in the will takes the form of contrary desires. One desire is the center of darkness and one is the center of brightness. The world is not contingent. But it is necessary for the self-revelation in time of God's eternity. Evil is the upsetting of order. Humankind is the link between the world and the angels. Human consciousness performs God's revelation in the world.

In *Forty Questions of the Soul*, Boehme described God as the *Urgrund*, something that is not dependent on any other being and which has no ontological determination. This is God's most fundamental aspect. Simultaneously, the *Urgrund* is complete and original freedom. No plurality, self-realization, or self-revelation is in the *Urgrund*. Instead, God has an eternal tendency to move in this direction.

Self-revelation has three elements: (1) the principle of consciousness, (2) the intermediary of self-objectification, and (3) the link between subject and object. In Trinitarian language, undefined will corresponds to the Father, will possessing identity corresponds to the Son, and will as self-revealing corresponds to the Holy Spirit. In addition, a fourth element, *Sophia*, exists: the divine wisdom that contains innumerable ideas that will be realized.

Boehme departed from earlier positions that presented God in spatial and temporal terms. For him God is separate from the world. God's nature has three qualities: (1) concentration, (2) expansion, and (3) synthesis that holds the first

two in tension. This is God's dark center. And it is the source of life and destruction.

Three qualities in which the spirit is made real compose light's center. The
first quality corresponds to love; the second corresponds to tone and contributes
to determination; and the third connects God with the world. A seventh quality,
fourth in this sequence, connects the center of darkness with the center of light.

God as good eternally defeats the potentially destructive power of non-being
that He contains. This victory translates into beings that arise by a dialectical
process. The world came into being by necessity. And it emanates from God's
nature. God's essence defines the world's essence. Hell and Heaven are eternal.
But the world is temporal and is subject to evil's prevalent influence.

Each created being is independent of the Creator because each being has in
itself its own ontological center received from God. Human beings possess
freedom. And freedom is a condition for moral responsibility.

Before Adam's fall, a fall occurred in the angelic world: Lucifer's. Lucifer
turned his imagination toward the center of darkness and started to want to
dominate light's center. Boehme thought that the existence of a world that
contains evil is better than no world at all. Furthermore, darkness is necessary for
light to appear. God manifests his love and grace by setting the world right.

Man is a microcosm and a microtheos. Thereby man mediates between God
and the world. Adam's fall was not the primary source of evil because Lucifer's
fall preceded it. Adam fell before he became a man of flesh and bone. (*Genesis*
presents the Fall occurring last.)

Adam had belonged to three basic principles. His role was to connect the
two centers of God. But Adam concentrated on himself, which separated him
from God. This led to a series of falls. Supposedly, *Genesis* describes the last fall,
where Adam opposed God and wanted to obtain the knowledge of "good and
evil."

Initially, Adam was an individual of both genders and could reproduce by
magic. When he became corporeal, Adam lost his intuitive and magical faculties.

Because a human being is a microtheos, he never lost his likeness to God.
A human being is also a microcosm. This means that despite the Fall, humankind
can play a part in rectifying evil.

A human being's unique position in the cosmos means that we have our in
own self-awareness and know God and the world. God also fully reveals himself
in human consciousness and language. The *vernunft* is reason, which abstractly
knows the world of phenomena. *Verstand* is intuition, which reaches the essences
of things. But fullness of knowledge is possible only by grace. When we truly
knows things we participate in our nature and we fully express that nature when
we pronounce a thing's name.

So an affinity appears to exist in Boehme's thought between human speech
and God's creative Word. When Adam gave names to things, he completed the
act of creation. In Adam's speech perfect agreement existed among a thing's
name, sound, and the essence. After the fall of the Tower of Babel, humankind

lost its natural speech and many different conventional languages arose. A few exceptional people can reach the nature of a thing by their own language. Speech also performs an important cosmic role because speech can connect the fallen cosmos with God. The way to salvation is to renounce our own selfishness or our tendency to focus our will upon ourselves, which was the cause of the Fall in the first place. When we concentrate on God, we each become a "microtheos."

For a long time many scholars regarded Boehme's writings as a product of pure inspiration, unadulterated by borrowings from others. This view facilitated the assimilation of his works. As Schelling wrote:

Boehme is a marvel in the history of mankind, and especially in the history of the German mind. If anyone could forget what a treasure of nature, depth of mind and heart there is in the German nation, he need only recall Boehme . . . .. Just as the mythologies and theogonies of primitive peoples anticipated science, so Boehme anticipates all the scientific systems of modern philosophy when he describes the births of God. Jacob Boehme was born in 1575, René Descartes in 1596. What was intuition and immediate nature in Boehme appears in Spinoza, who died 100 years after Boehme's birth, as fully developed rationalism.

Schelling here accurately describes how philosophers welcomed Boehme's views. Boehme's teachings were a kind of rationalization. Modern times have been a period for constructing philosophical systems. Rationalism does not consist in the interpretation of the rationality of being. It consists in constructing systems that propose models of reality. Contemporary intellectuals often call such a system an "ontology."

Rationalists seek their models in "sources" at the fringe of occult knowledge such as Hermeticism, the Cabala, mysticism, and theosophy. They do not seek them in reality. Boehme's work was one of these unofficial sources. Besides Eckhart, according to von Baader, Hegel's collaborator and publisher of Boehme's works, Boehme's doctrine had a strong influence upon Hegel. In addition, Boehme was a precursor of the theosophical movements of the late nineteenth and twentieth centuries.[19]

Part Six

THE TRANSFORMATION OF
PHILOSOPHY INTO IDEOLOGY

Lady Wisdom Morphing into Marx by Hugh McDonald

Twenty-Seven

IDEAS—CONCEPTS—IMPRESSIONS

The term "ideology" has a permanent home in modern languages. We associate it most often with politics and political parties as they advance their agendas. As we understand ideology today, it underlies political programs covering such important areas of social life as the form of government, economy, legal system, and education. The political program rests upon a conception of a human being and society and some vision of reality as a whole. An ideology addresses these matters as they concern the political realm.

Philosophy considers these same issues. So, we must adequately distinguish ideology and philosophy. Unlike classical philosophy, ideology implies a program for action aimed at social change. Social changes extend beyond any one nation. They can exist on an international scale. Ideology starts from some set of philosophical assumptions and seeks to subject all philosophy to itself. It seeks, especially, to subjugate the kind of philosophy that aims as a disinterested knowledge of truth.

How did ideology arise, and how did it work to transform philosophy into ideology?

While the etymology of the term "ideology" appears Ancient, Antoine-Louis-Claude Destutt de Tracy (1754–1836), a leading figure in the French Revolution, coined the term in 1776 from two Greek words: "*idea*" and "*logos*." The term signified a new approach to philosophy.

De Tracy was largely unknown and unstudied during the twentieth century. In the nineteenth century translations of his works appeared in many European languages. François-Pierre-Gonthier Maine de Biran, Thomas Jefferson, Augustine Thierry, August Comte, Karl Marx, the Dekabrists, coalmen, and different revolutionaries quoted him as an authority. Perhaps the contemporary ideologiza-tion of philosophy and civilization has proceeded so far that some intellectuals no longer know, or are reluctant to reveal, its genesis. [1]

Whatever the case, ideology was not initially, as its etymology might suggest, a science about Platonic ideas. When de Tracy wrote about ideas he looked to the philosophy of René Descartes, the British empiricists, and the French sensualists. In philosophical terms, de Tracy's ideology is a sensualistic offshoot of Cartesianism. And his transformation of Descartes's "*cogito ergo sum*" to "*je sens, donc j'existe*" ("I feel, therefore I exist") indicates his sensualism. [2]

The Greek noun *idea* comes from the verb "*horao*"—"I see." The term "*idea*" took on a philosophical meaning in Plato's thought. An idea is something that corresponds to our concepts (as distinct from sense impressions). And it is an immaterial exemplar cause for things in the material world. Plato thought that science would be impossible without such ideas.

Aristotle rejected Platonic ideas as an unnecessary hypostasis of concepts. In place of an idea, Aristotle presented a thing's internal form as the foundation of our general concepts. This alteration made scientific knowledge possible. Neo-Platonism located ideas in a separate Intellect, the first hypostasis emanating from the Primordial One.

The problem of ideas in God's Intellect arose among Christian thinkers when they considered creation of the world. Christians could not simply say that ideas or concepts were in God's mind in the same way as they were in the human mind. And they could not accept the existence of ideas as exemplars in the Platonic sense.

We associate human concepts or ideas with the way the human intellect interprets the world. We associate neo-Platonic ideas with the doctrine that beings appear by emanation. Meanwhile, in Christianity, God is the Creator. So ideas can only be different ways in which God beholds himself. Hence, early Christian thinkers could not conceive of ideas as being something "within" God's Intellect.[3]

Descartes completely changed how philosophers looked at ideas. Descartes maintained that an idea is a form of thought. Thought is that whereby we are conscious. Thoughts (*cogitationes*) include mental images of things, acts of will, and sentiments. Cartesian ideas are images of things such as a man, a chimera, Heaven, or an angel.

Cartesianism differed from the Scholastic tradition by regarding the idea as the immediate object of our knowledge. The Scholastic *medium quo* (a transparent intermediary) became the *medium quod* (that which we know). After Descartes, philosophers regarded the idea only as the internal correlate of our cognitive acts. As Descartes said in his *Meditations on First Philosophy*:

And that I may be enabled to examine this without interrupting the order of meditation I have proposed to myself, it is necessary at this stage to divide all my thoughts (*cogitationes*) into certain classes, and to consider in which of these classes truth and error are, strictly speaking, to be found. Of my thoughts some are, as it were, images of things, and to these alone properly belongs the name IDEA; as when I think [represent to my mind] a man, a chimera, the sky, an angel or God. Others, again, have certain other forms; as when I will, fear, affirm, or deny, I always, indeed, apprehend something as the object (*subiectum*) of my thought, but I also embrace in thought something more than the representation of the object; and of this class of thoughts some are called volitions or affections, and others judgments.[4]

Descartes presented mathematics as the model science and regarded as mechanical operations the functions of living organisms. These ideas would have a great effect on philosophy's direction. With respect to method, contemporary philosophers generally accept Descartes's principle that we should study the acts and faculties of knowledge before we try to know the knowledge's object.

Descartes intended this method to protect philosophers from errors. But its effect was to isolate human knowledge from real being.

When we put the mode of human knowledge before the object of knowledge, we will treat the object only as the intentional content of the act of knowledge. This intentional content is a possible, not a real, being. We can make no valid inference from what is possible to what actually is (*"a posse ad esse non valet illatio"*). As philosophers, we would no longer be analyzing reality. We would be occupied with contents already divorced from reality as our starting point.

British empiricism denied the entire spiritual sphere (Descartes' *res cogitans*) and focused upon the material (Descartes' *res extensa*). John Locke defined ideas as follows: "This term idea best signifies everything that is an object of thought; thus I used it to signify what we understand by an image of fantasy, a concept, a species, and everything with which the mind can occupy itself when it thinks."[5]

According to Locke, ideas have two sources: perceptions or reflection. Ideas are simple or composite; composite ideas result from the action of our mind and have their basis in some laws. Locke reduced philosophy's task to a study of the origin of ideas, how we possess knowledge by ideas, and what are the foundations of beliefs and opinions.

De Tracy says that he read Étienne Bonnot de Condillac before he read Locke.[6] Condillac had a strong influence on de Tracy.[7] Condillac argued against the role of internal experience and denied any action of the mind. He maintained that the content and functions of knowledge come only from impressions. Condillac was a radical sensualist. He treated the human mind as a vessel for impressions.[8]

Ideology is a system of thought that studies our powers of perception and the origin of ideas within us.[9] When de Tracy spoke of "ideas or impressions" and "the faculty of thought or perception," he was clearly identifying ideas with percep-tions, and the mind with the senses.[10]

The chief method of ideology is the analysis of the sense faculties and sense cognition to discover the first cognitive acts and the most basic components of the object of knowledge.[11] When we grasp these elementary units by observation and experience and they are simple impressions, then we can elaborate upon these units by deduction. Deduction is also a form of sense knowledge.[12]

In this perspective ideology must be the first and most important science, and the other sciences must be dependent upon it. If every science is a system of different ideas, then the first field must be the science that shows the sources of ideas, especially of fundamental ideas, and how we unite and divide ideas.

The first science must study the process whereby we form judgments to show which judgments are wrong and right. The same science must study our faculties of knowledge to determine whether we use them properly. Ideology is the science that performs these tasks. The other sciences simply apply ideology to their respective objects.[13] Because of ideology's special status, de Tracy described it as "the theory of theories" (*"la théorie des théories"*) and as the real first philosophy or first science ("*la véritable philosophie première ou science première*").[14]

According to de Tracy, ideas are impressions. No immaterial concepts and no souls exist. De Tracy thought that only ideas exist and that ideas are our whole being and existence.[15] For this reason, the study of ideas plays a fundamental role for all fields of knowledge.

His rejection of metaphysical questions related to his reservations concerning the term "metaphysics."[16] He was also suspicious of the term "psychology," which he thought implied the existence of the soul, which he held did not exist.

De Tracy needed a completely new term for the "science of impressions and ideas." Hence, the term "ideology." De Tracy used the term "metaphysics" as a synonym for ideology.

This change in terminology was a conscious effort on his part to imitate changes in nomenclature in the particular sciences, especially those introduced in chemistry (for example, by Antoine Laurent Lavoisier).[17]

De Tracy's entire plan for the division of the sciences was: The first place contains the means of knowledge, ideology as the science of the formation of our ideas; grammar as the science of the expression of our ideas; and logic as the science of joining and disjoining ideas. The second place involves the application of our means of knowledge to the study of the will and its effects: economy, which concerns our action; morality, which concerns our feelings; and governance, which concerns the direction of others. The third place involves the application of our means of knowledge to the study of beings other than ourselves: physics studies bodies and their properties; geometry studies the properties of extension; and calculus studies the properties of quantity.

De Tracy thought that we can reduce a complex idea to a simple idea that has its source in sense experience. If we cannot reduce an idea to a simple idea then we should banish it from science. Such an idea is equivocal or meaningless and judgments based upon it must be false.[18]

He based his conception of science upon the model of the physical sciences. One part of science concerns facts or experience. Another part concerns reasoning about facts. Likewise, two kinds of truth exist: (1) factual, and (2) deductive, or abstract truths.

If ideology, the initial and simple sense impressions, were the starting point for the humanistic sciences, he thought that the humanistic sciences would achieve the same certainty as the physical sciences.[19] De Tracy conceived of Newton's physics as the model for all science: its starting point is observation and experience, and a few physical laws expressed in mathematical form enable us to explain astronomical and mechanical phenomena.[20]

He thought that we could achieve the unity of all the sciences because the social sciences are the same in their method and structure as the material sciences. This followed from his radical materialism.[21]

De Tracy reduced the human mind to a physical object and thought to sense perception. His sensualism went farther than John Locke's because Locke admitted a difference between the senses and the intellect, even if this was only apparent in acts of reflection. De Tracy regarded acts of reflection merely as modifications of perception.

He maintained that four basic operations of the mind exist: sensation, memory, judgment, and desire. Even when he spoke of the human sciences, de Tracy thought that they concerned a physical, not a spiritual, being. Hence, De Tracy did not hesitate to say that ideology is a part of zoology.[22]

His doctrine belongs to the same line of sensualistic and materialistic monism represented by Julien Offray de La Mettrie, Étienne Bonnot de Condillac, and Paul Thiry d'Holbach. A human being is a being at the intersection of machine and beast, and society is a group of such animal-machines.

The change in direction in philosophy from metaphysical to epistemological questions, and from the philosophy of the object to the philosophy of the subject, made philosophers turn their attention to how human knowledge occurs. Ideology was one variation of the new understanding of philosophy and became increasingly dominant. Was the study of the mode of human knowledge based on mere curiosity, or were other motives present in the case of ideology?

Twenty-Eight

THE WAR AGAINST IDOLS

Classical philosophers saw the dangers associated with cognitive errors. They sought the sources of these errors and their remedies. They most frequently attacked credibility of sense knowledge, although some philosophers in the Skeptical tradition argued against the validity of intellectual knowledge.

René Descartes and Francis Bacon took a new approach to the problem of errors. In Descartes's system, skepticism was a method for finding an indubitable starting point . In Bacon's system, knowledge of the source of errors and of the different kinds of errors was a condition for reaching true knowledge. Bacon called these universal errors idols, which are innate or acquired.[1]

Bacon lists four general kinds of errors: Idols of the (1) tribe, (2) cave, (3) market, and (4) theater. Idols of the tribe are rooted in human nature when human reason transfers its own properties to things. Idols of the cave happen when we distort our vision of things by our own nature, upbringing, contact with other, books, or authority. Idols of the market are errors contained in the language we hear from others. Idols of the theater are errors rooted in philosophical systems and in wrong rules of proof.[2]

Any successful acquisition of true knowledge and all intellectual progress depend upon knowing how to recognize errors. Errors are more than private and accidental matters. They are universal and are linked with the structure of human knowledge, or are inscribed in the cultural tradition. Ignorance of errors introduces falsehood into our quest for knowledge and hinders scientific advancement.

With respect to the conception of science, Bacon's rejection of authorities, philosophical systems, and traditional methods of seeking knowledge is crucial. If Bacon's critics, and even some of his successors, did not regard him as quite an original thinker, they were in agreement that he played an important role in undermining the entire heritage of previous philosophy and science. Bacon, especially, had called for a break with tradition and the past. This had a greater influence upon the history of philosophy and culture than did the new theory of induction, in which Bacon did not significantly depart from Aristotle.

Beyond the question of idols in the search for knowledge, Bacon maintained that philosophers considered the boundaries of legitimate knowledge. They considered the limits and scope of scientific knowledge, the transgression of which would be a departure from science.

As John Locke wrote:

> If by this study of the nature of the mind I will be able to discover its abilities and its range and to state to what things these abilities are in some measure suited, and where they deceive me, then I think that the results may

be helpful in disposing man's restless mind to consider when it is grappling with things that exceed its ability to conceive, to hold it back when it touches the furthest limits of its reach; to make it calmly come to terms with its ignorance about the things which upon investigation turn out to be inaccessible to our mental powers.[3]

In Locke's approach traditional metaphysical and theological questions necessarily lose their place as a part of the search for scientific knowledge. Philosophers could reject the heritage of the Greek philosophers and Christian theologians, not because it was untrue, but because it was not scientific according to the new conception of science.

What Bacon called idols, the major types of cognitive errors, other philosophers such as Étienne Bonnot de Condillac, Claude-Adrien Helvetius, and Paul Thiry d'Holbach called prejudices.[4] They saw these prejudices as obstacles that had hindered scientific and civilizational development. And they thought that the major source of these prejudices was religion, especially Catholicism.

The modern conception of science and philosophy is more than one among many variations or models of science. It has become a weapon for fighting the Ancient conception of science, classical philosophy, and religion, all in the name of science. When we treat science as a weapon without regard to the truth of its propositions, it becomes a part of ideology.

Twenty-Nine

THE CRITIQUE OF
CATHOLICISM AND METAPHYSICS

When Antoine-Louis-Claude Destutt de Tracy established ideology, he saw one of its tasks as the removal of prejudices that come from metaphysics and religion. He regarded the Catholic Church as the main stronghold of these prejudices. The battle against the Church in the French Revolution was more that a struggle for political power or influence: a struggle for the foundations of civilization. The revolutionaries wanted to strike at the type of civilization the Church represented and nurtured. By saying that religion and metaphysics were unscientific, they disqualified those areas of culture from having any part in society, and they wanted especially to disqualify the Church.

De Tracy thought of science as the technique of discovering knowledge's sources, degree of certainty, and limits. Metaphysics could no longer be a science because, according to him, metaphysics speaks of principles and an end for all things, and it teaches that God is the source and destination of all things. Those questions belong to the art of imagination, not to science. Their purpose is to give a feeling of satisfaction, not to provide information.[1]

The old metaphysics was to the new as alchemy was to chemistry and astrology to astronomy. Ideology became the remedy for the illness of traditional metaphysics, and, simultaneously, became the new metaphysics. De Tracy spoke interchangeably of "ideology or metaphysics" so that we would recognize ideology as holding the first position among the sciences, just as metaphysics had been first philosophy in classical philosophy.

Some metaphysical theses that de Tracy criticized included the claims that general ideas give meaning to particular ideas, the mental world is separate from the material world, and human beings move only in a world of pure thought.[2] He set forth a scientific epistemology (or ideology) based upon experience and sense impressions as opposed to metaphysics, which he maintained accepts *a priori* presuppositions as its starting point. He regarded Francis Bacon, Thomas Hobbes, John Locke, Helvetius, and Étienne Bonnot de Condillac as representatives of scientific epistemology, and Plato, Aristotle, René Descartes, Gottrfied Wilhelm von Leibniz, George Berkeley, and Immanuel Kant as representatives of metaphysics.

De Tracy went farther than Condillac in his opposition to metaphysics and religion because he eliminated theology as a science.[3] He regarded theology as an infantile philosophy based upon imagination, not reason.[4] He maintained that religion's source was the primitive mind's terror of invisible forces, which it conceived as an explanation for otherwise inexplicable phenomena. Because the primitive mind did not know the true causes of these phenomena, it looked for explanations in metaphysics and theology. The priestly caste gave these fields scientific status and so hindered the development of true science.[5] They gave to purely imaginary being the status of divine spiritual beings and ascribed to them different fictitious properties. Christian dogmas were nonsense expressed in a language of sophistry. Christ was not God. He was one among a thousand versions of the Sun-god. Christ had no more historical validity than Hercules, Bacchus, and Osiris.[6] Consequently, science should help to demystify religion.

Increasingly evident is that the modern conception of science as ideology is directed against religion, metaphysics, and theology. The new science intends to expose these fields as false and illusory, to battle them, and strip of them of their status in society.

Thirty

IDEOLOGY AND THE NEW SOCIAL ORDER

The planned remedy for human knowledge aimed at going farther than improving our grasp of the truth in a defined area. It aimed beyond the development of those domains that had acquired the status of sciences. Its concern was with more than the domination of nature. It aimed at improving us and society.

The new conception of science built upon ideology concerned morality and politics. Personal happiness and a perfect society became the new ends for philosophy and ideology's intended aims. The utopian thought that appeared at the start of the sixteenth century in literature was transformed into a philosophy supported by ideology. Utopia became a political program in the framework of the new conception of science and philosophy. On this account, Antoine-Louis-Claude Destutt de Tracy plays a crucial role in philosophy's history.

Some sections of science that de Tracy developed reflect the connection between ideology and moral and political life. He tried to show in the form of equations the following sequences of identities: the faculty of thought or perception=knowledge=truth=virtue=happiness=the feeling of love=freedom =equality=philanthropy.[1]

We move from ideology to personal happiness, and then to a well-organized society. De Tracy also followed John Locke and Étienne Bonnot de Condillac in thinking that we can make morality and politics an object of deduction.

We can explain these quite startling arguments and equations with a kind of ethical and political intellectualism: the body of errors passing for knowledge that comes from metaphysics or religion is the source of unhappiness; true knowledge is the source of personal and social happiness. On this account, ideology has a scientific, philosophical, and social dimension (its primary dimension). Ideology prepares the way for humankind's happiness. And this is its aim.

A new morality and a new politics grew out of, or were deductions from, ideology. Education and the legislative system introduced the new order.[2] De Tracy was especially involved in educational reform. He supported the revolutionary tendencies to eliminate the influence of the Church in education. He wanted ideology taught before grammar and logic.

De Tracy's ideology is not the same as the kind of ideology we have in mind today when we speak of Communist or Liberal ideology. His ideology was an epistemology and concerned how ideas or impressions appear in our cognitive faculties. It was also more than an epistemology or philosophy of human knowledge because it had broader philosophical and practical ends. It would cleanse philosophy of metaphysics. And he intended it to be the basis for a new social order.

This social order was a repudiation of the classical tradition with no place any longer for the human being conceived as a person (since the human being is merely a reservoir of impressions) and no place for any reference to God. De Tracy subordinated his ideology to these broader goals. And they later became a part of ideology, which, as de Tracy intended, has ideological ends such as we understand them today.

Thirty-One

IDEOLOGY:
THE EVOLUTION OF THE CONCEPT

As the opposition of "true" science and philosophy to the old philosophy and religion intensified, the term "ideology" started to find a place for itself. Karl Marx treated ideology as a false consciousness of a historical character that revealed social contradictions. He maintained that religion was its primary source.[1]

Marx thought that social conditions contain a contradiction that consists in class differences and class struggle between the dominant and the subordinate classes. A failure to find a practical resolution causes a sublimation or projection at the conscious level.[2] Ideology as a projection serves as a justification for the ruling class's dominance. And religion supports this justification.

Marx wrote of ideology in a pejorative sense because it solidified existing social injustice at the conscious level. Vladymir Ilyich Lenin (1870–1924) thought that ideology concerns every social class, not only the dominant class, because ideology includes the broad interests of every class. So a bourgeois ideology and a proletarian ideology exist. In this sense, "ideology" is not directly a pejorative term. But it has a pejorative sense with respect to some interests and specific classes.[3]

Other Marxists (such as György Lukács, Antonio Gramsci, Karl Mannheim, and Lucien Goldman) referred to ideology in terms of social class. They maintained that revolution and science, by which they meant the so-called "scientific world-view," could resolve social contradictions.[4]

Ideology also appears in modern psychology. Sigmund Freud thought that ideology initially concerned an internal psychological state that contained a contradiction, and that we need to resolve this contradiction externally (which was the reverse of Marx' position).[5]

Wilfredo Pareto, in turn, thought that irrational social mechanisms formed an important part of ideology, and that ideology served as a rationalization of these mechanisms. In this sense, we cannot eradicate ideology from social life, because people need a rational explanation for their irrational behaviors. Ideology is a necessary tool of all political authority, which, in its governance of society, has to present proper motives for a specific set of social behaviors.[6]

Émile Durkheim returned to Francis Bacon's theory of idols to explain ideology. He thought that ideology contains all the ideas or idols that stand between the scientist and the facts. Ideology is a mental illusion. Durkheim initially thought that the illusion had only an individual character. Later, he maintained that it had a social dimension. Still later, Durkheim no longer saw science in opposition to ideology because he thought that both came from the same source, which is shared social experience.[7]

The concept of ideology also made an appearance in the human sciences, especially in psychology and sociology. In these contexts new elements occurred to Antoine-Louis-Claude de Tracy's original conception. One constant element was that ideology still attacked tradition, metaphysics, and religion, to the extent that those domains affected social life. In modern time the possibility of gaining such influence became the most important motive of life and prestige.

Ideology criticizes religion and metaphysics to appropriate their role in social life. Ideology has a sacral, fundamentalistic, element. Ideology treats its tenets (such as class struggle in Communist ideology, or the free market in Liberal ideology) like infallible dogmas. Ideology also has its own undisputed metaphysical doctrines, such as the doctrine that God does not exist, or that the human being is a product of evolution. Ideology took the form of a knife dipped in poison that cut into the basic forms of social life and very often perverted familial, political, and economic life. Without a sane metaphysics and revealed religion, no way exists to expose ideology's false foundations, presuppositions, methods, and activities.

Part Seven

TOWARD A NEW WORLD ORDER

The New World Order by Hugh McDonald

Thirty-Two

KANT: THE WORLD
BEFORE REASON'S TRIBUNAL

The leading question in Immanuel Kant's scientific syestem was no longer why things happen and exist. It was to identify the *a priori* conditions for events. In a descriptive sense, this was the epistemological ground for the question of "how" that appeared with Positivism. In asking "why," the scientist addresses things and seeks their internal and external causes and principles of being. Even Plato's idealism was ultimately an attempt to answer the question of the causes of the world that we perceive through our senses.

In Kant's case, the question addresses the process of knowing, not things themselves. Kant conceived the knowing process as a relation between subject and object. The question of *a priori* conditions arose because he was unable to find anything in a phenomenon that could meet the conditions for its being a scientific object: necessity and universality.

Kant then summoned knowledge itself before pure reason's tribunal. The result was his critique of pure reason.

The critique aimed to show the limits of human knowledge, and, in the case of what we know scientifically, to show the *a priori* conditions that make this knowledge possible. The tribunal of the critique of pure reason showed that the scientifically valid contents of knowledge come from the subject, not the object. For this reason the scientific question cannot concern things. It must address *a priori* conditions, what comes from the subject, for only that contains scientifically valid contents.

In the order of sense knowledge, Kant maintains that space and time are *a priori* categories. In the order of rational knowledge, Kant mentions twelve kinds of judgment and twelve corresponding categories. He draws these judgments and categories from formal logic.[1] The *a priori* conditions belong to the knowing subject. But they have a dimension beyond the individual and belong to the human race.

Surprising to see how a legal mindset influenced Kant's model of scientific knowledge. Francis Bacon spoke of torturing nature to give up her secrets and compared this to the testimony of tortured slaves in Roman law. Kant called reason before the tribunal similar to a summons before a law court:

[I]t is a summons directed to the reason to resume its most onerous task, to know itself and so that the tribunal can decide, which may vindicate it in its just demands or reject all groundless claims not by arbitrary resolutions but on the basis of eternal and unchanging laws; the critique of the pure reason is such a tribunal.[2]

Pure reason is also the source of the critique of pure reason. Pure reason summons itself to the tribunal to show that scientifically valid contents come from pure reason, while things in themselves are outside the range of knowledge. In effect, what we call the world or nature comes from the subject. Kant says that we generally interchange two expressions: "world" and "nature."

He maintains that "world" as world "denotes the mathematical total of all phenomena and the totality of their synthesis—in its progress by means of composition, as well as by division." We direct our attention to an aggregation in space and time to think about the world as a quantity.

In contrast, when we use the term "world" to refer to nature, we regard the world "as a dynamical whole." We direct our attention to the "unity in the existence of the phenomena," not to "the aggregation in space and time, for the purpose of cogitating it as a quantity."[3]

Kant claims that the world is an *a priori* synthesis of impressions with the categories of space and time, while nature is the world as we conceive it in terms of dynamism. Neither nature nor the world is real. They are merely phenomenal, and their content comes from reason.

Kant performs a transvaluation of the conception of "reality." He regards assertoric judgments as true and their affirmation or denial as real. He calls denial or affirmation in apodictic judgments necessary. And he says that denial or assertion in problematic judgments is merely possible.[4] In this way the concept of reality depends on the modal state of a judgment. This, in turn, depends *a priori* upon the intellect. What comes from the subject, what is *a priori*, is most real.

Kant maintains that, in a thing, or, more precisely, in an object-phenomenon, we scientifically know only that what we put into it. Knowledge is more a recognition of contents that the subject projects than knowledge of something different from the subject. Not surprising that in Kant cognition moves from empirical consciousness (which mix with impressions) toward pure consciousness, which considers only its own *a priori* categories. Knowledge's ideal is a flight from the real world!

As Kant says: "Now, a gradual transition from empirical consciousness to pure consciousness is possible, inasmuch as the real in this consciousness entirely vanishes, and there remains a merely formal consciousness (*a priori*) of the manifold in time and space." And, "Now, a gradual transition from empirical consciousness to pure consciousness is possible, inasmuch as the real in this consciousness entirely vanishes, and there remains a merely formal consciousness (*a priori*) of the manifold in time and space."[5] Kant reduces the existence of a being to its being known by transcendental reason.

Mieczysław Albert Krąpiec comments that, through this reduction, Kant radically roots the existence of objects in the subject. He writes:

Thus even the existence of objects ("existence" and "object" obviously as Kant conceived these) remains radically rooted in the subject. The subject demarcates the objective "reality" that is understood in this way. Thus everything with which human knowledge is concerned does not take place in relation to the things or to the really existing world, but is "the process of becoming aware" by the construction of the objects of the knowledge of what occurs within me when I know.[6]

Kant makes a distinction in the knowing subject between the empirical "I" and the transcendental "I." The empirical "I" is what we can know when we objectify it. But we cannot objectify the transcendental "I." "[A]s regards internal intuition," he says, " we cognize our own subject only as phenomenon, and not as it is in itself." "[A]s regards internal intuition," he adds "we cognize our own subject only as phenomenon, and not as it is in itself." Kant writes further: "The consciousness of self is thus very far from a knowledge of self, in which I do not use the categories, whereby I cogitate an object, by means of the conjunction of the manifold in one apperception."[7]

Simultaneously, through the intellect the function that *a priori* connects the variety of impressions into a unity manifests itself. Kant speaks of the unity of apperception. It consists in a synthesis of different data in one intuition. This, in turn, accounts for the stable identity of self-knowledge.[8] We cannot know the transcendental "I." And we cannot apply the category of substance to it. This category is only an *a priori* category referring to the empirical world.

Regarding the intellect's *a priori* function, Krąpiec remarks:

> In the margin we should note that the lack of content in the "self" should not lead us to deny the "self" as a substance. What is essential is that the "self" always appears as the performer and the existing subject who is "without content," a subject who exists on his own, while his acts are acts and emanations of the "self" that is affirmed directly and without reference to content. . . . The self as subject is a subsistent being, namely a substance.[9]

Just as our *a priori* projections come to the foreground in cognition, so the moral order is not something that we find to which we must then relate. Morality is not the result of understanding any real good. Practical reason projects the moral order. Practical reason's Kantian postulate ("Act only on that maxim whereby you can at the same time will that it should become a universal law.") sounds sublime and even resembles the evangelical counsel, "Do unto others as you would have them do unto you." Yet it is an *a priori*, purely formal, postulate containing no content.[10]

Most importantly, it makes no mention of any good. At best the good may appear as something secondary. We are dealing here with a dictate of the reason that sets ethics as a theory-of-morality before morality considered in itself as the order of the good.

From the moral viewpoint, this attitude is a sign of great pride. Difficult otherwise to recall any situation in which we dictate to reality what is right and wrong. Kant's explanation of duty (deontologism) apart from our normal desire or happiness testifies to a pride based upon an unhealthy egotism.[11] The situation becomes even worse when such a dictate falls into the hands of the State. Then we must deal with totalitarianism.

We do not solve the problem of morality by introducing values as the correlates of obligations (*sollen*). Values do not belong to the real order. And values do not intersect with goods.

For Kant obligation is no completion of the order of being. The two are opposed (the opposition of *sein* and *sollen*). In classical ethics, we actualize or realize a potency for the sake of an act, end, or good. We understand reality as it is incompletely actual, or in potency. We understand the good by analogy. And we are guided in our action by the good that should be done (*bonum est faciendum*).

In the particular sciences the consequences of Kantianism lead to a civilization dominated by technology. Kant had great enthusiasm for modern physics, especially Newton's physics. Newton's physics provided him with a model for the epistemological foundations of his theory of scientific knowledge. When we summon the world before the tribunal of reason it must then give an account before technology. As Martin Heidegger writes:

Fittingly, man's accustomed behavior finds expression in the emergence of modern natural science: the kind of representation proper to him is imposed on nature as on a calculable system of forces.

Modern physics is experimental physics not because it approaches the examination of nature with an apparatus, but on the contrary, because physics now as a pure theory arranges nature so that *a priori* it presents itself as a calculable system of forces. The experiment is planned to ask whether nature has so arranged accounts for itself and how it appears. . .. The modern physical theory of nature paves the way first not for technology, but for the essence of contemporary technology.[12]

Hard to disagree with this diagnosis. Unfortunately Heidegger's essay does not point to Kant as the thinker who first, in theory, called nature to give an account before the reason. Technology's growth already made possible our ability to summon nature in a practical way. And nature was increasingly unable to defend itself from such regimentation and experiments.

We can look at Kant's whole conception from an epistemological viewpoint, But we may also take a metaphysical viewpoint. When we examine Kant's epistemology in metaphysical terms, his position resembles a kind of pantheism: no individual subject exists; a general, or universal, human self exists common to all people.

A projective creativity in the order of knowledge characterizes this universal self. So, too, in science, the order of action, and reality. This self is like the divine self.

Kant was an heir to the gnosticism prevalent everywhere during his time. Kant's successors such as Georg Wilhelm Friedrich Hegel and Max Scheler said what Kant had left unsaid, but for which he laid the foundations. Reality is a divinity that creates its own metamorphoses. Reality's core is the self in which at least some elect have a share.

Directly, positively, Max Scheler attributes pan-en-theism to Kant:

The idea and subsistent value, the value of existence (*Seinswert*) of the human person which moreoever is superior to all possible utilitarian and life-related values, can be accepted only by someone who together with Kant and all the other great European philosophers sees in man a citizen of *two* worlds and also, as we gladly say, he sees man as a being rooted in two different and essential attributes of one, substantial, and divine principle of the world.[13]

Kant's system laid the groundwork for the contemporary change in the scientific question from "why?" to "how?" If reality is unknowable then: what can we know? If the real good is beyond my reach, what should we do? If the immortality of the soul is only a postulate of the practical reason, for what can we hope? If we do not understand different aspects of reality in the light of causes, we are compelled to seek fixed rules of knowledge and conduct and laws that show how things are composed and operate. Knowledge of such rules and laws lead our search for knowledge in science toward utilitarian goals, for which the normative question of science is chiefly "how?", not "why."

Thirty-Three

AUGUSTE COMTE: TOWARD A NEW AGE

Positivism was a continuation of the conception of science initiated by Francis Bacon and Descartes. The demands that Positivism placed on science were even more rigorous than in the period of the scientific revolution. The Positivistic conception of science claimed to be the most scientific approach, untainted by any elements of mythology, religion, or metaphysics. Positivism embraced a paradigm of science as opposed to domains such as metaphysics and theology that pretended to be scientific. Although the English term "science" comes from the Latin *scientia*, Positivism restricts its meaning to modern physical science.

Auguste Comte (1798–1857) founded Positivism. Comte's version of Positivism was not limited to a changing specific conception of science or philosophy. It aimed at changing civilization as a whole. While the conception of science was essential, it was only a part of the undertaking.

Comte mentions six different senses of the term "positive", which is at the root of the term "Positivism": (1) the real, not the illusory; (2) the beneficial, not the futile; (3) the certain, not uncertain; (4) the precise as distinct from value; (5) the opposite of the negative, what serves to build, not destroy; (6) the relative, not the absolute.[1]

Comte's ultimate aim was to lead humanity on the road of progress. Positivism's motto was "order and progress."[2] On the basis of this motto, we can see clearly that Positivism as chiefly an enterprise of civilization, not science.

The growth of industry and sociology were important parts of Comte's plan. Industry would supply humanity with increasing means of living. Sociology would perfect humanity.[3]

Sociology was Comte's response to the social changes for which utopian literature had called. Comte thought that with social laws sociology would discover we would be able "to perfect the turns of our fate."[4] His conscious aim was to achieve a state of social perfection and happiness "without God and King."[5]

Comte thought that human knowledge progresses through three phases: (1) theological, (2) metaphysical, and (3) positive. He described the theological phase as "fictional." In the theological phase we try to study the internal nature of things by looking for first causes and final causes. We look to supernatural causes, and we seek absolute knowledge.[6]

He also calls the metaphysical phase "abstract." In this stage we replace the supernatural forces of the theological phase with abstract forces, beings capable of causing all phenomena.[7] Comte says that only stage we may call "scientific" is the positive phase:

Finally in the positive phase we recognize that it is impossible to obtain absolute concepts; we renounce the investigation of the origin and destination of the universe and the quest for knowledge of the internal causes of phenomena; instead we try to discover their laws, which means their relations, sequences, and similarities, and we use reasoning together with observation. The explanation of facts becomes merely the establishment of connections between particular phenomena and certain general facts whose number tends to decrease due to the progress of science.[8]

Positivism's scientific character consists in its rejection of metaphysical questions and its concentration on acts whose relations we understand with the help of laws.

Comte's theory of science makes frequent mention of Francis Bacon and René Descartes. For example, he says,

As a result of our entire intellectual evolution, especially the evolution of the movement that from the times of Bacon and Descartes has taken in all Western Europe, a normal state of mentality must finally set in after such a long wandering along pathless ways, because the positive philosophy includes the whole extent of investigations and will acquire the rational foundations it still lacks.[9]

He emphasizes that we base true science upon the observation of facts. He eliminates search for first and final causes from science.

Instead, science investigates the circumstances in which phenomena appear. It links the phenomena with relations of sequence and similarity.[10] Science is concerned with more than fact-gathering. Science primarily seeks to discover laws.[11]

These laws should extend to all phenomena, inorganic and organic, physical and moral, individual and social. When we know these laws, we can predict events. Science's laws provide more than ordinary erudition.

Knowledge of these laws has crucial practical consequences in social life.[12] According to Comte, the law of gravity is a universal law whereby we can explain all the phenomena in the universe.[13]

Comte mentioned the following benefits of the Positive philosophy. (1) It is a rational means that emphasizes the logical laws of intelligence. (2) It provides direction for a complete transformation of the system of education. (3) It contributes to progress in different sciences. (4) It serves as the basis for social reorganization.[14]

Comte's presentation of the structure of science distinguishes theory and practice. When he speaks of theory he does not have in mind science in the classical sense of *theoria*. He means science's theoretical aspect that has a practical end: "In considering human work as a whole we should conceive of nature as intended to provide foundations for human action. In short, we express this thus: *from knowledge to prediction, from prediction to action*; this formula expresses the relation of *knowledge* to *practical skill*."[15]

Science has a practical end and an essential reference to time and changing circumstances, for it connects with practical action. Comte says, "To see in order to foresee is the essence of science."[16] In the framework of science's utilitarian end (*propter uti*) another specification appears: to see (to know) so as to foresee (*scire propter providere*).

Because Comte wants to see satisfied "the needs our mind feels when it desires to know the laws that rule phenomena," he says that we cannot reduce science to practical needs alone.[17] We should not let these words mislead us. Comte is not suggesting a return to the *theoria* of Ancient times, to knowledge for the sake of knowledge.

Comte's point is that, when our mind is interested in more than practical matters, it is more fertile and, in effect, provides more possibilities for practical applications. Therefore, he writes, "[T]he most important applications arise out of theories that were formulated out of a purely scientific intention and developed over centuries without gaining any practical results."[18] When Comte speaks of a theory he always thinks of a theory evaluated according to its possible practical benefits.

Comte thinks that theory and practice need each other, that practice without theory would limit thought's progress. And theory without practice would prevent us from completing action.[19] He maintains that (1) purely theoretical ambitions result from the "mystical and absolute character of primitive theories" and (2) contempt for practical thought taints these ambitions. Theories of this kind do not contribute to progress.[20]

The structure of science matches the structure of professional life: a scholar works on theory, a director directs the work of others, and engineers at the middle level have theoretical and practical knowledge and skills.[21]

Comte distinguishes two ends of science: (1) concrete and descriptive and (2) directive. The natural sciences are concrete and descriptive. They apply the discovered laws to the history of specific kinds of things.[22] Comte lists the following sciences in the framework of Positive philosophy: astronomy, physics, chemistry, biology, and sociology.[23] And he says that mathematics is the foundation of all philosophy.[24]

Nonetheless, Comte regards sociology as the leading science. Just as mathematics underlies the sciences, sociology directs the sciences.[25] Mathematics is the start, sociology the end.[26] We cannot separate sociology from politics because politics is the practical translation of sociology's findings.

Positivism shows that only relative, no absolute, points of view exist. And they depend upon sociology.[27] Sociology studies society as a true reality, while individual people are only abstractions.[28] Only humanity can achieve the desire for immortality, the individual cannot.[29]

Morality must also be subordinate to a social point of view.[30] Sociology joins theory with action.[31] Mathematics leads to excessive theorization.[32] Industry

has a goes beyond these with its own crucial role in linking theory and practice.[33]
Science's methods include deductions and induction. We should employ deduction in specialized research and induction to discover general laws.[34] By releasing philosophy from theology and metaphysics, humanity can move forward on the road of progress.

Comte does not restrict himself to the critique of theology and philosophy. He offers practical prescriptions for eliminating these domains from social life; we must eliminate the ministerial theological and metaphysical budget and then liquidate the Chairs of theology and metaphysics. A more refined method is a systematic silence on problems of theology and metaphysics.

Today, many universities throughout the world universally apply this method of using administrative tools to fight metaphysics and the humanistic sciences that stem from the classical heritage.[35] In this way, they complete the work that Bacon and Descartes initiated.[36]

Comte inherited the idea of the progress of humanity from the Renaissance. Jean-Jacques Rousseau, Jean D'Alembert, Claude-Adrien Helvetius, Denis Diderot, Anne Robert Jacques Turgot, Marie Jean Antoine Nicolas Caritat Marquis de Condorcet, and others had propagated this idea.[37] Comte's critique of purely theoretical knowledge in the name of the greatest benefit also was not new. D'Alembert said that, while we must be mindful that some accomplishments of science with no application at present will someday have use, useless knowledge is unnecessary.[38]

François Marie Arouet de Voltaire (Voltaire) and Diderot criticized purely contemplative knowledge and promoted the idea that a human being is primarily a *homo faber* (man the maker). Simultaneously, Comte gave Positivism the status of a global civilization. Positivism became a new religion, a new ideology, humanity's highest priest and deity. Comte identified humanity with the Supreme Being. So, he completely subordinated the individual person to society as to his god.[39]

Since society is a god, the science of society, sociology, becomes a religion and a kind of divine omniscience. Comte recognized Christianity as a historical and sociological fact, but treated Jesus Christ as a fiction.[40] In this way, he destroyed the true God in a sociological sense. Humanity took the place of the true God, which means that sociology became *sociolatria*, the adoration of society.[41]

Scholars tried relatively early to distinguish two different phases in Comte's work: (1) his works on Positivism (1830–1842); (2) his discussions of the Religion of Humanity (1851–1854). Scientific circles accepted the works of his Positivism, but rejected those concerning the Religion of Humanity. Some scholars connected Comte's thought in the second phase with a brain disease.[42]

My question is: Is the Positivistic conception of science truly independent of Comte's vision of Positivism as a civilization and religion? Does Positivism add such a new dimension to the conception of science? Apparently, the conception of science becomes a tool for making changes to civilization. The cult of science does not spring from a love of science or truth. Ideology motivates it. A conception of science provides a weapon to attack the foundations of Western civilization. And these foundations include religion and metaphysics. The purpose of the attack is to establish a completely new civilization.

Thirty-Four

NEOPOSITIVISM AT WAR
WITH METAPHYSICS

Scolars customarily divide Positivism into three phases. The first phase comprises the professional careeer of Positivism's founder, Auguste Comte. The second phase is the empirio-critical phase in the late nineteenth century as represented by Richard Avenarius (1843–1896) and Ernst Mach (1838–1916). Moritz Schlick (1882–1936) led the third phase, that of the Vienna Circle. Schlick held a chair at the University of Vienna after Mach, Rudolf Carnap (1891–1970), and Ludwig Wittgenstein (1889–1951).

These schools shared a minimalistic conception of science and a negative attitude toward metaphysics. In some sense we may even say that the Positivist and new conception of science has a negative purpose: to eliminate metaphysics from among the sciences.

Comte tried to eliminate metaphysics by changing the guiding question in science from inquiry about causes to descriptions that do not refer to causes. This allowed him to banish metaphysical questions from science.

Followers of Comte such as Mach and Avenarius held that "how," not "why," is the question that should guide science. Science answers the question "how?" by resorting to "pure experience" expressed with the help of "pure description."[1]

Pure experience is the foundation of science and contains no elements borrowed from ordinary, or naïve, experience. It contains no elements that are the result of human projection. Pure experience concerns only facts, which we can describe, not explain. Our description must be economical. We must express as simply as possible and must include as many facts as possible. Such a description of facts provides the basis for formulating general laws, although each fact occurs only once and is different from all other facts. Laws are abbreviated accounts of facts.

In science a single concept replaces many mental images, and a single law replaces many particular statements.[2] The purpose Comte assigned is unchanged: the task of science is to enable the prediction of the sequential occurrence of facts to gain the greatest number of practical benefits.

Easy to see that the Neopositivist conception of science borrowed some elements from the nominalist tradition and British empiricism. Nominalism emphasizes the individuality of facts. Empiricism focuses on an exposition of the role of experience and how we progress from experience to concepts and laws. This approach in science goes farther than eliminating those metaphysical questions regarded as false: it eliminates all metaphysical questions, which it regards as being useless for the purposes of scientific research.

Hanna Buczyńska-Garewicz maintains:

The term metaphysics was applied to all knowledge that was not a description of facts and which went beyond a description of facts. The ambition of the critical empiricists was the development of that conception of science and the liquidation of metaphysics. They wanted to liberate science from metaphysics. Critical empiricism was strongly connected with the development at that time of the natural sciences, and with its anti-metaphysical tendency it created a favorable climate for the development of modern knowledge. In the period when a series of generalizations and non-empirical and vague concepts were being eliminated from the natural sciences as being useless ballast, the philosophy of "pure experience" was a reflection of this state of affairs and, at the same time, gave support to these processes.[3]

In this view, only the natural sciences meet the criteria for science and provide positive knowledge about the world.[4] The natural sciences alone provide the paradigm for all science.

Logic and empirical verifiability are specific features of scientific knowledge. The Neopositivists identified the meaning of a proposition with the principles of its verification: only verifiable propositions are meaningful. In logical verification the Neopositivist considers the structure of propositions: empirical verifiablity depends on the degree to which experimental methods have advanced.[5]

We may verify scientific propositions by comparison with facts. They are descriptive propositions. The theses of metaphysics are not subject to empirical verification. So, they have no scientific status.

The Neopositivists also regard metaphysical propositions as unscientific and logically meaningless. Meaningless expressions such as "being in itself" and "the question of all being" appear in such propositions. We cannot empirically verify the meaning of such expressions.

Metaphysical propositions also break the rules of logic that govern the construction of propositions. An example would be a metaphysical proposition in which "is" occurs in a metaphysical sense. The Neopositivists held that the verb "is" can only occur in a structure of predication and that use of "is" in a proposition without a predicate violates the rules of language.[6]

In this way the Neopositivists dealt a death-blow to any metaphysics that speaks of a being as something that exists: real being, which can only be being as that which exists, must vanish from our picture of the world. Hard to imagine a more radical kind of idealism than the elimination of existence from being.

Neopositivist methodology appears to allow elimination of metaphysics as unscientific and meaningless. Carnap writes: "The supposed propositions of metaphysics, the philosophy of values, and ethics, are merely expressions of feelings that in turn evoke in the listener feelings and dispositions of the will."[7]

Such being the case, the theses of metaphysics have only an emotive and volitional, but no cognitive, meaning. Logic sets the boundaries of scientific knowledge, and of the rest we must be silent (Wittgenstein).[8] The only place the Neopositivists see for philosophy is in the analysis of scientific knowledge. But philosophy has no role in the description or explanation of the world.

In this way, after the Cartesian revolution, philosophy became (1) divorced from reality; then (2) directed to the contents of consciousness or ideas; later (3) to the process of cognition. Finally, (4), in Neopositivism, philosophy became the study of the particular sciences and their languages. In Neopositivism philosophy was cut off (1) from reality to move at the level of meta-language, and (2) from all metaphysical questions.

Critics of Neopositivism most often mention the many metaphysical propositions it contains. Neopositivism is a hidden metaphysics. It cannot escape from metaphysics. The major objection to the Neopositivist method is that, in itself, the way it defines the field and method of scientific knowledge is non-scientific.

We cannot establish a general method of scientific knowledge unless we understand the scientific object. While escargots and mushrooms are edible, we must gather them by different methods. The object of mathematics is different from the object of metaphysics. The investigation of numbers requires one method, the investigation of being as being another.[9]

Neither the Neopositivist concept of "pure experience" nor the definition of the object philosophy as the analysis of the language of science is a pure description expressed in mathematical language. According to its own criteria for science, Neopositivism is non-scientific. If Neopositivism is not science, should we regard it as a project, a prophecy, a dictate?

Because of the subtlety of its objections, this demasking of Neopositivism has not had a wide affect in society. Neopositivists simply reject the objection that their doctrine is a crypto-metaphysics. They treat Positivism as a religion and an ideology and hold fast to their conception of science. Their conception of science has aims beyond science: it starts with practical concerns and ends with an aversion to Christianity.

During the twentieth century, many intellectuals regarded the Positivist and Neopositivist conception of science as effective. And it started to dominate science and civilization in general. To this day Positivism in science and civilization is an obligatory canon of the Western world. Even today the West uses Positivist assumptions to measure scientific knowledge, scientific progress, and the accomplishments of so-called "humanity."

Positivism goes further by setting goals for social action. Postmodernism is an attempt to break the model of Positivism. It has had a greater influence on the humanities than on the mathematical and technical sciences or on the industrial and technological Easternization of our civilization.

If the Positivist and Neopositivist model of knowledge and civilization is too narrow, if the Positivist paradigm contributed to the rejection of our inheritance of Western culture wherein we loved and esteemed *theoria* and the love of theoretical knowledge affected the sciences and all parts of culture, we may ask, Is Positivism really capable of filling the place left by a true, rational, religion, by theology, metaphysics, philosophy, contemplation, or personalism?

When Positivism denies classical culture entry into our civilization, it finds itself unable to control the side-doors. Sects enter in the place of the Church, and theosophy enters in the place of theology. Cultivation of mythical thinking replaces philosophy. Totalitarianism replaces liberty. The spread of theosophy and Eastern thought among the elite and the spread of sects at every level of society have been a reaction to Positivism's austerity. Yet hardheaded Positivists and mystical Easternists share a contempt for our classical inheritance.

No surprise that the Positivist mentality opposed to metaphysics and Christianity could so easily accommodate Eastern thought and practice. We should keep this in mind when we analyze the phenomenon of Positivism and its one-sided tendency. Many Positivists readily succumbed to forms of mysticism far from the ideals of Positivism. Even the standard-bearers of Positivism, such as Comte and Wittgenstein fell into forms of pseudo-mysticism. We cannot regard this simply as a personal aberration. It must be a consequence of a line of thought that leads to absurdity at the level of civilization. One response is a flight to mysticism.

The heritage of Positivism has lingered in the West to this day. Encyclopedias perpetuate it as a standard. Textbooks present it as the basis for understanding science and as the key to the proper interpretation of history, human beings, and society. Nevertheless, imagining a conception of science more steeped in ideology than Positivism is difficult.

Thirty-Five

SCIENCE AND THE NEW WORLD ORDER
OR
SCIENCE IN THE SERVICE OF GLOBALISM

Science has existed as a distinct realm of culture in the Western world for two and a half millennia. Only in modern times have people changed civilization in the name of science. In the name of some conception of science, people have changed science itself, morality and mores, art, industry, and religion. Today, we popularly conceive a civilized person to be someone whose actions follow scientific principles and who sees humankind's only salvation in science.

Bertrand Russell thought that, about 1650, science became the dominant factor in shaping the opinions of educated people, whereas technological economy arose around 1850. In that relatively short period, he maintained, the sciences have exerted an extraordinary influence upon human life.

> Man has existed for about a million years. He has possessed writing for about 6,000 years, agriculture somewhat longer, but perhaps not much longer. Science, as a dominant factor in determining the beliefs of educated men, has existed for about 300 years; as a source of economic technique, for about 150 years. In this brief period it has proved itself an incredibly powerful revolutionary force.[1]

Only in the twentieth century did the possibility arise to use science to transform the whole world and all humanity. Science has become ubiquitous. All departments of culture have gradually become "scientific" to the point where they no longer have distinct identities. Technology first enabled us to move from manual craftsmanship to industrial methods. It is now leading us into the post-industrial era. When we speak today of a post-industrial society, we think primarily of a new stage in the development of science and its influence on the shape of civilization. Highly developed technologies led by electronics have become more important than industry as they change the face of the world.

The growth of science first intended to serve the growth of industry, which would supposedly lead to the conquest of the world in keeping with the Protestant and capitalist principle of constructing an Earthly paradise.[2] Communism also took up this mission, but without success.

The current stage in scientific development allows increasing control over nature and society on a global scale. For some time now we have heard of a "scientific society." Russell said, "I call a society 'scientific' in the degree to which scientific knowledge, and technique based upon that knowledge, affects its daily life, its economics, and its political organization."[3]

As Zbigniew Brzeziński has observed, "The post-industrial society is becoming a 'technetronic society': a society that is shaped culturally, psychologically, socially, and economically by the impact of technology and electronics—particularly in the area of computers and communications."[4]

A single, central authority would have to rule such a society. That authority would hold the reins for ruling society. It would control government, law, finance, the media, the judicial system, and education. Likely that such authority, if it became possible, would have to maintain itself by force, not by democratic consent, at least in its first phase. As Russell says, "It seems to follow that a world government could only be kept in being by force, not by the spontaneous loyalty that now inspires a nation at war." With the development of effective social techniques for forming public opinion, democratic procedures may be extend, because the regime in power will control consciousness.[5]

Movements to create a single world government to direct the matters of all humanity are nothing new. Some thinkers proposed such a government during the late Middle Ages. Dante Alighieri was among the first late Medieval, early Renaissance, Western thinkers to advance the idea of one world government.

Dante gave many arguments to support his idea that one common government would best serve humankind. He based most of them upon a quasi-syllogistic deduction that he derived from a univocal and abstract concept of humanity. This showed the influence of Averroes on him. Dante reasoned that just as human reason directs our faculties and parts, the father directs the family, the ruler directs the city, and the monarch directs the state, one supreme emperor should govern humankind for its own good. Such governments resemble most of all God's rule, for one God rules the whole universe.[6]

Dante's mistake was that he separated the social context of human life from culture. We differentiate cultures by nations, if only for historical reasons. We cannot reduce politics to a matter of administration, because the end of politics is the common good, which culture, not administration, realizes. Use of an excessively abstract notion of humanity as the proper subject of politics can also have dangerous results, such as totalitarianism.

Only at the end of the twentieth century had such a program become a real possibility. In modern times (for example, with Abbé de Saint-Pierre Charles-Irenée Castel—1713), one motivation behind the idea of calling into existence a world-wide government was the need to limit war and ensure a universal and lasting peace. Russell presented a similar motivation, "Since war is likely to become more destructive of human life than it has been in recent centuries, unification under a single government is probably necessary unless we are to acquiesce to either a return to barbarism or the extinction of the human race."[7]

We popularly call this pursuit of one world government "globalism." Globalism has possessed a clearly socialistic mark since the start of the nineteenth century with thinkers like Claude Henri Saint-Simon.[8]

Political globalism aims to usher in a "New World Order." President George Herbert Walker Bush was the first politician officially to use the expression "New World Order." In a speech to the American nation in 1990, he said: "Out of these troubled times, our fifth objective—a new world order—can emerge."[9] Globalism initially encountered two sets of obstacles, technical and political. Without the proper means of execution, plans for world domination would remain dreams.

Modern history shows us many States or absolute monarchies that have accumulated power and struggled for domination and wider spheres of influence. Over time ideology took the place of the interests of the monarchies and States. In the twentieth century, some ideologies, such as Fascism and Nazism, confined themselves to a single State or people, while ideologies such as Communism, socialism, and Liberalism reached beyond the confines of any one state.

The globalist tendencies in the latter ideologies found a greater base of support in international structures than in the interests of particular states. Science played an important role in many ways in all these ideologies. Each ideology tried to justify itself by an appeal to science.

Fascism, Nazism, and Liberalism claim their basis in science. Nazism appealed to a theory of race. Communism to a theory of class struggle. All these ideologies regarded religion as their chief enemy. So, they could not base their claims to legitimacy on supernatural causes. They fought against religion in the name of reason and science. And they looked to science for their primary support.

Science became the primary tool, handmaiden, of ideology in its pursuit of its goals. Science gave, and expanded, power. Scientific investigations of the cosmos, the Earth, nature, society, and human beings would provide the key to domination. Rivalry between ideologies was largely a rivalry in the field of science.

After World War II, the world politically divided into two spheres of influence: the United States representing Liberal ideology, and the Union of Soviet Socialist Republics (U.S.S.R.) representing Communist ideology. In the early 1990s, the U.S.S.R. officially dissolved. The new Russia would not be Communist in its constitution. The world lay open to the expansion of Liberal ideology.

In the meantime Liberal ideology took on more elements of the socialist ideology. Social democracy is based ideologically upon so-called "social liberalism" and has increasingly become the foundation of "globalism".

We should make a distinction between the kind of socialism based upon defined extra-national organizational structures that have a definite political profile, cosmopolitanism (which is an expression of an attitude of openness to the world and other peoples), and Catholicism (which is organizationally open to all humanity, but under the aspect of religion, not in the political arena conceived as a play of forces).

The Soviet Union realized only one kind of socialism. The official announcement that the Soviet State had fallen did not yet mean the end of world-wide socialism. Within the Soviet Union a struggle had existed between Josef Stalin who proclaimed the primacy of the Soviet State, and Leon Trotsky who considered world revolution to be the highest priority. Michail Gorbachev, the main architect of "*perestroika*" (restructuring), explained the meaning of the changes: "The essence of '*perestroika*' is to connect socialism with democracy and to revive the Leninist conception of socialism, both in theory and in practice . . . we want more socialism." Gorbachev has lived for several years in the U.S.A. and still belongs to a close circle of globalists from around the world.[10]

With the collapse of the Soviet Union, the political and ideological battle for world domination, in some sense, finished. But socialism in its Western version had not lost. Of the 15 States of the European Union, in the year 2000 all but one had socialist governments.

Liberal Democrats and Liberal Republicans in the United States are the American counterpart of the European socialists. With the help of parts of the American media, they have tended for decades to set the tone for political life in the United States. Increasingly clear, though hard to believe, is how far the socialist agenda has recently advanced in the United States. A crucial point to note since, as Brzeziński has said, "After the defeat of its rival, the United States found itself in the unusual situation of being the first and, at the same time, only truly world-wide power."[11]

Globalization would have no chance of success without science. The development of science, especially in the twentieth century, made possible actions in many different areas on a huge and unprecedented scale. Economy, finance, energy, communication, the military, education, and the media all have an increasingly global reach because of the advance of science.

Stanisław Wielgus maintains that modern science and technology are developing increasingly quickly. He states that institutes of scientific development in the world say (1) every minute scientists discover a new chemical formula; (2) every three minutes they discover a new physical compound; (3) and every five minutes they make a new medical discovery; (4) nine of ten natural scientists who ever lived since the start of known civilization are alive and work now; (5) in the U.S., the number of scientific researchers doubles every thirty years; (6) we will print more books in the next few decades that were printed in rom the invention of the printing press until now; and (7) the number of personal computers in the world doubles every eighteen months.

Past centuries prepared the ideological ground for globalism's reign in the humanistic fields, such as philosophy, psychology, and sociology. The twentieth century saw unprecedented growth in the particular sciences. The particular sciences had a utilitarian outlook and provided technologies whose effects local barriers did not impede.[13]

Science plays a role in making performance of some actions technically possible and in planning the scale of actions. In the past, utopian writers dreamed of plans that exceeded the humanly possible. Today's socialists are re-working such plans according to science today and are bringing them to life.

Some fields have developed with the help of science to such a degree that effective action in them is now possible on a global scale, and other fields approach this level. Social life in modern times has been the domain of independent States. Such States include government, administration, the legal system, treasury, armed forces, judiciary, and education. These spheres are becoming increasingly internationalized and will ultimately be globalized. Science enables institutions such as the U.N., NATO, the World Bank, and the International Court to operate on a global scale. Without science these institutions would be mere facades of power.

The change in the guiding question of science from "knowing why" to "know-how" is certainly a source of this tremendous scientific growth. Science in the service of utility has assumed monstrous forms and posed a threat to the natural environment over the entire planet. Human beings are now the target of genetic engineering, and we now use science in social engineering to affect entire societies. Science has provided the key to torturing nature, and now provides the key to torturing society and human beings.

Tendencies toward the scientific control of society were characteristic of the twentieth century. Political authorities and business interests have used science successfully to achieve their goals. Scientists examine human nature and society in laboratories and use the findings for purposes of political power and profit.

Scientists study the human genetic code, physiology, emotions, the conscious mind, the subconscious, the will, and the intellect. They break down what they study into primary elements. Their methodology of "know-how" prompts them to make some practical use of their findings. So they use their science to control human beings. Consequently, they tend to treat the human person and interpersonal relations as mere objects.

This objectification has perilous consequences. It results practically in a kind of cognition of human beings that inevitably instrumentalizes us. This kind of cognition is know-how. The question "how?" has an instrumental character. If the question concerns human beings, then we treat human beings as objects in the research process. Worse, the findings of such research provide unimaginable means for dominating humankind.

Science certainly can benefit human beings. But, in the new scientific approach, before we can speak of benefits or dangers, part of the scientist's starting point treats a person as a tool, because the conception of science builds its self-conception on the question of know-how. The question of know-how turns us into a thing as an object of research. Often the benefits of science become a screen to hide science's use for sinister ends. One example is genetic engineering. Scientists kill countless human embryos in the name of improving the genetic code. We may also mention the application of science in euthanasia, abortion, and biological weapons.

Scientists carefully study society, especially mass reactions. Business and politics use their findings to direct mass populations and human resources at the level of the State and at the level of international and supranational organizations.

And, today, science provides the principles and directions to achieve these ends. The ingenious or charismatic leader has become an anachronism. Today, we package some people through a manufactured image designed to satisfy the psychological desires of the general population, who need a personification of political authority. Because of the cult of democracy and democratic procedures such as voting, those who seek political power today must use modern science.

These people use scientific methods from statistics, psychology, and sociology in public opinion research polls and political campaigns. A serious politician or political part today needs a team of scientists. They are more than advisors. They make diagnoses and prescribe the most effective strategies. Specialized firms that use the services of scientists today develop selective strategies for businesses and politicians.

Large business enterprises today require knowledge of public reaction. They must know how to influence the public effectively. In large firms the effects of advertisements, layout of buildings, placement of merchandise, and behavior of employees pass through the filter of science to ensure successful action and profit.

Science's findings (science meaning "know-how" concerning human beings) increasingly become tools that can be used against human beings. A tool's essence is that we can use it for opposite purposes. The materialistic conception of a human being dominates civilization today. It denies us the status of persons that transcend nature and society. So the ultimate end of political power becomes the domination over us under all aspects.

This materialism emphasizes our social value. Although this sounds sublime, it can lead to a situation where politicians treat us as no more than an element of society lacking any distinct personal identity or any avenue to transcendence. Socialism has become the dominant political ideology. And, in its pure and original form, it conceives of human beings only in terms of society.

We derive the term "society" from the Latin *socius*. Pierre Leroux (1797–1871) authored the term "socialism." He saw in this new ideology the final realization of social and religious goals. At its roots socialism is a secular or naturalistic form of religion.[14]

This conception of "social" differs from the Ancient Greek conception of human beings as social animals. The Greeks saw that we need society to live and develop as people. But each of us is still a distinct being. In the socialist conception, society, not the individual human being, is the primary subject.

The high level of development in science makes domination of human beings a possible element of society. The science at its starting point takes away any chance for us to ask the ultimate questions about our origin and end. This kind of science only teaches us how to use things. Despite its many accomplishments and the useful goods it teaches us to make, in the final analysis, today's science will serves our social, familial, and personal enslavement.

Before our eyes we see the construction of the utopia of which philosophers and novelists accurately and prophetically warned us. George Orwell is the best known of these novelists. Not without some irony, some writers note that Orwell's *1984*, written in 1948, concerned events that had already started in 1930 in the United States. By 1978, 100 of 137 of Orwell's pronouncements had been fulfilled.[15] The intense surveillance described by Orwell continually expands over the entire world. And it has its basis in the latest scientific advances.

Official public presentaton of the biochip occurred on 16 October 2000, in New York. A biochip under a person's skin can coordinate with the Global Positioning System (GPS). In the near future, the biochip will be a means to monitor particular individuals to an unprecedented degree. The code of the biochip will allow officials to access all the information gathered on a particular person in a special government computer.

In short, physical science in the twenty-first century is increasingly becoming an element of a global civilization that levels cultural differences as it lowers the level of the our life as spiritual and sovereign beings.

Part Eight

THE PLACE OF SCIENCE IN CULTURE

Lady Wisdom Overseeing The Prime Mover by Hugh McDonald

Thirty-Six

WHAT IS CULTURE?

The word "culture" comes from the Latin *cultura*. It originally meant the cultivation of a farmer's field. Hence, we still speak today of agriculture.

In his *Tusculan Disputations* [*Disputationes Tusculanae*], Marcus Tullius Cicero compared a human being's upbringing to the cultivation of the land. Just as we need to prepare and tend the land to bear fruit, so Cicero said a human being needs the right cultivation to acquire virtues and eliminate vices. He spoke of the culture of the soul, *cultura animi*.

> *Nam ut agri non omnes frugiferi sunt qui coluntur, falsumque illud Accii: Probae etsi in segetem sunt deteriorem datae Fruges, tamen ipsae suapte natura enitent, sic animi non omnes culti fructum ferunt. Atque, ut in eodem simili verser, ut ager quamvis fertilis sine cultura fructuosus esse non potest, sic sine doctrina animus; ita est utraque res sine altera debilis. Cultura autem animi philosophia est; haec extrahit vitia radicitus et praeparat animos ad satus accipiendos eaque mandat eis et, ut ita dicam, serit, quae adulta fructus uberrimos ferant.*

(And just as fields, to use the same metaphor, cannot be fruitful without cultivation [*cultura*], so the soul cannot be fruitful without learning. The first thing without the second is thus feeble. The cultivation of the soul is philosophy; it pulls out vices by the roots and prepares minds to accept seed, and commands them, so to speak, to sow in order to bear the most abundant fruit when the plants have matured.)[1]

In the classical conception, culture unites at its foundations with nature, *physis*. Culture should complete or perfect nature.

Human nature operates through different faculties. We see with our eyes, hear with our ears, and walk with our legs. We learn to walk and think. Left to themselves, these operations and their corresponding organs fail to develop. They degenerate.

Human nature differs from the nature of beasts in that beasts possess instinct. For the most part, instinct infallibly guides their particular organs to maturity to reach their optimal state. We possess more than instinct: reason, a faculty joined with the immaterial soul and the subject. Reason must direct human life and the development of our different faculties.

In the classical sense, culture completes what is lacking in, and perfects, nature. Reason guides culture. Culture is the rational completion of what nature lacks. Reason has a necessary role in culture because the other faculties by themselves cannot perform optimally. Because we possess more than instinct, only reason enables us to see our end. Reason drives human nature as we use it to select human nature's proper means properly arranged in time.

Culture's subject and end is the human person. Even if we speak of cultivating a field, we cultivate it for the sake of nourishing people, not for the sake of the field itself. In the strict sense, culture as the cultivation of the soul is an upbringing that corrects vices and inculcates virtues. Culture in this sense disposes a person to be able to direct his or her life toward the authentic human good. Culture as upbringing aims at the human good in the perspective of our whole life and in view of human life's ultimate end.

Instinct such as we have it arises spontaneously and needs little help from parents. But human reason takes a time long to develop before it can serve as a reliable guide. Parents are not the only ones who educate their child's reason. Schools, the Church, and the heritage of past generations also contribute. Our individual efforts and immediate milieu are inadequate to develop the potentialities and openness of human reason.

Cultivation by reason, where reason cultivates the subject, is crucial in our lives. In reality, in some way, everything in culture cultivates reason and, thereby, cultivates the human being. Human rational cognition appears here in different functions and forms, only one of which is science.

Immanuel Kant once wrote:

We possess two expressions, world and nature, which are generally interchanged. The first denotes the mathematical total of all phenomena and the totality of their synthesis—in its progress by means of composition, as well as by division. And the world is termed nature, when it is regarded as a dynamical whole- when our attention is not directed to the aggregation in space and time, for the purpose of cogitating it as a quantity, but to the unity in the existence of phenomena.[2]

The post-Kantian conception of culture is diametrically opposed to the classical conception I have just presented. In the post-Kantian version, culture is no completion of nature. Nature is only a phenomenon created by an *a priori* synthesis of impressions with the categories of the intellect.

In this conception of nature, culture becomes an autonomous sphere in which we create values.

In the classical sense the human good was the rational completion of human nature. The human good came from the human soul, was an act, and directed toward human nature as a determinate potentiality. The axiological approach of Kant's system presents the human being as the authentic co-creator of culture. In so doing, it regards the human being as a god.

While worship of the human being and of values sounds quite noble, in reality, it leads to a false relation to the world, the things we produce, and ourselves. It robs nature of any value as a reality. Nature becomes merely raw material for technology to process. Nature treated as the correlate of the particular sciences appears only as matter unveiled in atomic structure (electrons, positrons, neutrinos, protons, and so on). Matter conceived in this way is "cold," in itself valueless.[3] Nature's only value is as raw material, and matter in itself has no value.

When technology develops to the point where it virtually eliminates barriers of space and time, it poses a threat to nature on a planetary scale. Such a view regards the Earth as nothing more than raw material that technology can arbitrarily rework. For this reason, while many of us today praise the human ingenuity behind technology's triumph, simultaneously we fear a series of ecological disasters on an unimaginable scale. Military and industrial technology could destroy the whole Earth.

Technology's triumph is a tribute to human ingenuity. But it does not necessarily coincide with an authentic respect for the human being as a person, as distinct and individual substance. It reduces human nature to a phenomenon, a synthesis of impressions and the intellect's *a priori* categories. In this sense, inasmuch as human nature is part of nature, we may subject human nature to technical procedures.

Genetic engineering has reached the point where cloning and transgenics are possible. These are grave dangers to humankind and individual people. Such possibilities rob newly conceived human life of personal dignity, and treat the components of the human body as interchangeable with parts from other animals and plants. These are consequences of a mistaken conception of our human nature and human body.[4]

The cult of humanity that Kant propagated in his famous maxim, that humanity should be treated only as an end, not as a means, and his idea of human dignity, do not protect the real human individual from the technical invasion of the ontological structure, of which human nature is an essential part. When we treat humanity and dignity in separation from nature, they become high-sounding, but empty, phrases. Culture in the post-Kantian sense poses a real threat to the world of nature and to human beings. The threat comes from a false vision of the world and the human person. We must rationally connect culture with human nature. Only then can human action turn to the real good.

Thirty-Seven

THE DOMAINS OF CULTURE

In a tradition that goes back to Aristotle, three primary domains of culture exist: (1) *theoria*, (2) *praxis*, and (3) *poiesis*. *Religio* was adds a fourth domain. But the Greeks treated religion as a part of moral action, *praxis*. Reason plays a crucial role in all three domains. *Theoria* is *recta ratio speculabilium*—right reason about theoretical matters. *Praxis* is *recta ratio agibilium*—right reason about things we can do. *Poiesis* is *recta ratio factibilium*—right reason about things we can make.

Poiesis considers making things. Traditionally, it includes the imitative arts, later called the fine arts, and industry, manual arts, servile arts, or craftsmanship. Other cognitive powers, such as the senses and imagination, and the emotions play an essential role in artistic creativity. Reason directs the creative process, but does not supplant the other faculties.

If we understand this relationship between reason and the other faculties we can avoid falling into the traps of poetic intellectualism and irrationalism. The creative process is not the work of the intellect or senses alone. Many faculties work together under the guidance of reason to realize an idea or vision.

Realization of such a vision is the aim of creative production. The aim of the fine arts is to delight the viewer. The aim in industry or craftsmanship is to achieve a proper functionality.[1]

Praxis includes the entire field of morality: (1) personal morality, which ethics studies; (2) family morality, which economics studies; and (3) social morality, which politics studies. Ethics considers personal good. Economics considers familial good. Politics considers social good. Each field requires the proper direction of reason. But thought alone cannot replace actions the other faculties perform.

Morality requires reason to acquire skill in finding effective and honest means to the end. Reason also shows the other faculties what is their proper measure. Justice is an acquired disposition of the will inclining the will to render another what is due. Fortitude and temperance are acquired emotional dispositions inclining us to hold to the so-called golden mean. In prudence reason must learn to direct us in matters of morality. In the other virtues reason should show the other faculties where good and evil are, so we may seek good and avoid evil.

God is the main point of reference in religion. When we perform acts of religion, we engage all our powers, inferior and superior, intellect and will, imagination, emotions, and senses. Since God is immaterial and infinite, our immaterial faculties (will and intellect) are central in religion. In the act of faith the will and the reason work together in different ways, depending upon which aspect of faith we emphasize. A unique problem because God is transcendent to us in terms of being and knowledge. While we may strongly engage and express our feelings, the act of faith finds its basis primarily in spiritual acts, not feelings. In view of God's essence and our human nature, spirituality should be the basis for personal relations between us and God, because the spiritual sphere is characteristic of personal beings. From the formal viewpoint, faith is an act of the reason, which acknowledges the revealed truth as truth by virtue of the will. As St. Thomas Aquinas states, "*[C]redere est actus intellectus, secundum quod movetur a voluntate ad assentiendum.*" ("To believe is an intellectual act according to which the will moves the intellect to assent.")[2]

Theoria takes in the whole sphere of knowledge called science. It is an acquired disposition and skill of reason to know the truth about reality for the sake of the truth. When, at the start, we reject truth as the end of theoretical knowledge, we virtually destroy this, and every other, domain of culture. While particular scientific disciplines may remain, they will belong to some other domain: *praxis*, *poiesis*, or *religio*; not to *theoria*.

This has been the recurrent situation in the history of Western culture. When we subordinate science to ethics or politics, art or technology, religion or mythology, science departs from truth as its proper end, and presents a distorted picture of reality. Then appears a univocal conception of science conditioned by the particular field to which we subordinate science. This conception then serves to support a paradigm of science. And, in the name of the new paradigm, we start to reject formerly recognized sciences as unscientific, and we elevate other disciplines to take their place.

If science does not belong to its own cultural domain and has no primary connection with truth, then, if it appears at all, it will bear *a priori* presuppositions from some other domain of culture and will have some end other than truth. If we develop science within the framework of *theoria*, we should understand science analogically, not univocally. Science is analogical because science's object is analogical.

Being, or reality, is the first object of scientific knowledge. Science is about knowing the truth about what is real. And reality is analogical. Knowledge, therefore, must be analogical if, indeed, it addresses reality. To know the truth for its own sake about being as being, about any of being's categories, or some aspect of being, we must apply different methods of knowing fitted to a different material object (what we are studying) and a different formal object (the aspect we are studying). This analogical variety reflects being's structure. Since we want to know truth for its own sake, we will know the truth about different categories of being in different ways.

Scientific knowledge's analogical character also reflects human knowledge's varied modes. Human knowing is varied and works through cognitive powers that perform different functions. A crucial point, because our mode of knowledge is inadequate to the way things exist. All the more, variety can exist in scientific knowledge.

While many types of scientific knowledge exist, we cannot suppose that no common definition of science exists or that we apply the designation of science arbitrarily. Scientific knowledge must have some differences that set it apart from ordinary knowledge, or from knowledge as it occurs in the other domains of culture.

To qualify as scientific knowledge, a discipline must have its own object and method of investigation and justification fitted to that object, and it must have an end. The findings must be a set of ordered and rationally justified propositions.[3]

Thirty-Eight

ENDS, LIMITS, AND DIRECTIONS
IN THE GROWTH OF SCIENCE

The development of science in the framework of *theoria* raised human knowledge in worth and responded to our natural desire for knowledge Aristotle recognized this when he wrote the first line of the *Metaphysics*: "*Pantes anthropoi tou eidenai oregantai physei.*" ("All human beings by nature desire to know.")[1] As persons, human beings, possess reason. Seeking knowledge is reason's fundamental and natural, or proper, operation. In a broader perspective this is also the completion of our humanity. In this way the truth actualizes our cognitive potencies. For the intellect, to live is to know, and to recognize that what we know agrees with reality is to know the truth. This agreement may occur spontaneously, accidentally, or from supernatural sources. It may be a reflective and methodical process, in which case it can become science.

While many cultures and all of us, by our natural desire to know, esteem truth's nobility, or worth, not all cultural domains have simultaneously developed in equal measure, in every civilization. Initial development of science, *theoria*, desire to know for its own sake, as a distinct cultural domain, is peculiar to Western culture. The historical fact is, it started here. And the Ancient Greeks created it.

If *theoria* has no distinct role in a particular culture, this does not mean that the culture is not a human society, that it is not advanced, or that it lives only on falsehood. When I speak of *theoria*, I am talking about a cultural domain, a determinate, theoretical way of knowing, a consideration of truth as truth.

We should readily recognize theoretical fragments from other cultures that agree with some strictly theoretical scientific assertions. In varying degrees, theoretical human reason ultimately makes itself known in every cultural domain. But rational fragments of knowledge are not the same as science cultivated in the framework of *theoria* as a distinct cultural domain, or as part of a cooperative cultural effort.

Human reason is potentially open to infinite knowledge. But, in our human condition, we acquire knowledge progressively, with difficulty, and over a long time. Human knowledge develops in a social context. Its history extends over many generations and nations. As a cultural domain, science is certainly a good for all humankind because it is a response to a natural human desire.

At the outset, not everyone understands science's greatness and importance. And not every culture esteems science. This is because not all people are aware of what science is or have a correct understanding of it. The ability to see science's true value and all that goes with it and the level of true education testify to a society's level of development.

Science's ultimate state, as developed within the framework of *theoria*, is the truth that enables us to progress. Human reason is an essential part of our humanity. We are persons, not mere things. Truth actualizes our reason, causes it to exist in act. As part of *theoria*, science's concern is with truth as truth. In this sense, science is the highest human guarantee of the truth. Science must be verifiable. We must rationally justify its theses. And science must be intersubjectively communicable, a communal effort. We cannot formulate science in a private or enigmatic system of signs.

If science remains in the framework of *theoria*, it is in no danger of becoming an absolute or being treated like a god. Such science is not an end in itself. Its end is truth. But truth is not isolated from reality or from the knowing subject.

Science's ultimate aim is to enable us to learn and understand reality because then we can live in the most human way. Science helps us actualize ourselves in our most personal actions.

As *theoria*, science allows us to see science's limits with respect to the object known and knowing subject. Science does not provide the complete truth about reality. And it cannot fully satisfy our human intellectual desires. Theoretical science must be modest in its claims. But it remains precious.

Science shows us that, while something is within the range of our desire for knowledge, science and human effort do not make it knowable as such. When theoretical science modestly and frankly acknowledges its own limits, then we can see that a rational place exists for a supernatural order of knowledge that is more-than-scientific.

Theoretical science can also show us what is the instrumental approach to science and the dangers it brings. When we divorce science from *theoria* and reduce it to a question of "know-how," science leads to a dangerous instrumentalization of us as a real human subjects, and of truth as science's end.

When this happens the human subject and science's end become means to something else. In this way our picture of reality becomes distorted, with ill effects for all the other domains of culture, including morality, art, and religion. The human subject as the bearer of dignity plays an essential role in morality. We have the natural and moral right to seek a real human end. Practically considered, instrumentalization of human nature in the service of practical science reduces us to amoral agents by inverting the end of a free human nature to serve scientific power, technocracy.

The denial of our natural human end of *theoria* robs human action and human nature of any intrinsic meaning and replaces it with instrumental meaning. The need to sublimate lower states of sensual knowledge and emotional states vanishes from art. By art we should have a more personal experience of such states.

Instrumentalization affects religion by attacking the first two commandments. This constitutes its most perfidious and perverse attack. It turns the human subject and the infinite, Absolute Good, into means, tools. The paradigm of utilitarian science attacks theoretical science. As a result, culture loses its ability to engage in rational self-reflection.

In his work *O ludzką politykę* [*On a Human Politics*], Mieczysław Albert Krąpiec shows how an incorrect conception of science, based upon *a priori* notions of the object and philosophical method, an *a priori* accepted theory of evolution, and an *a priori* notion of anthropology, poses a serious threat to the proper understanding of human nature and human rights. Mistaken *a priori* notions contribute to a deformed image of human nature and make defense of real human rights impossible.[2]

Only by a connection with *theoria* can utilitarian science provide authentic service, not destruction, to human life. Devoid its natural subordination to *theoria*, utilitarianism separates science from morality and religion and makes it their enemy. We then use science to manipulate or destroy people and to wage war against God. Actually, this effect winds up anti-utilitarian because it serves no real, even practical, human purpose. And science becomes an instrument of mass-destruction.

In his encyclicals, especially *Evangelium Vitae* (*The Gospel of Life*), and *Solicitudo rei socialis* (*Concern for Social Reality*) and in his homilies, Pope John Paul II wrote much on the dangers associated with scientific development. His homilies also consider the role of the university in the modern world.[3]

To understand the role and importance of science in culture, we must see its sources, limits, and end. The cult of science today is a facade for an ideology that is destroying the world, culture, and humankind. This ideology negates the personal dimension of human life while, simultaneously, it proclaims a program of atheism and wages war against God in different ways. By reflecting more deeply upon science we can escape manipulation, seek the truth, and seek help in and through science.

In this way, we can avoid becoming what Plato called "misologues"—people who distrust and despise all rational arguments because we have only seen them in the form of sophistry. Science has many names. But science in the chief sense, the primary analogue for all the different types of science, is theoretical science, which helps us to understand reality. Science in the form of technology is secondary to *theoria* and constantly connects with it. Technology separated from *theoria* becomes part of the domain of *poiesis*, making things. Then it may pose a threat to culture and human beings. *Theoria* is science's primary paradigm and the foundation of a truly scientific culture.

Thirty-Nine

CONCLUSION

Two and a half millennia ago, Aristotle said that the human race lives by art (*techne*) and reasoning (*logismois*). When we look around today we see confirmation of his statement. We live in a world in which science, especially in its practical technological applications, plays a dominant role.

Science in culture has gone through many phases since Aristotle. In Greek times, science, *theoria*, had truth as its end. Contemplation was science's crowning point.

During Hellenistic times, Stoics subordinated science to ethics and politics. Neo-Platonists taught that love and ecstasy were higher than rational knowledge, and magic could replace science.

During the Latin Middle Ages, Christian thinkers worked at a synthesis to show how faith and reason agree. They tried to preserve the Greek ideal of truth as a treasure and as the crowning point of human life. The Christians taught that we achieve this highest point of contemplation in the beatific vision, we directly behold God.

Finally, in modern times, under some Eastern influences transmitted to us through the Renaissance, a utilitarian approach in science became dominant. This tendency continues today.

Technology's triumphs are undeniably impressive. So are its dangerous failures related our spiritual life. The consumer life-style lacks higher purposes. Hence, it unwittingly uses social technology to dominate mass media, violently lowers educational standards, and brings the primacy of technology into higher education. All these cultural disorders result from the contemporary primacy of utilitarian science within the West.

Utilitarian science now occupies center-stage in Western culture. From this vantage point, it uses scientific methods to destroy science as a kind of wisdom and culture's noblest domains.

These changes in the face of civilization are astounding, terrifying. Never before has humankind been in such serious danger as today when we our own creation, technological science, threatens us.

Apart from miraculous intervention, our main hope to save ourselves from the devastating effects of technological science lies in restoring to science its the proper natural hierarchy of ends and means, subject and object. Treating a human being as a utilitarian object is wrong because we are real subjects, persons. We cannot treat even technology's most perfect products as ends because, by their essence, they are only means.

Every civilization's proper, primary, aim is to assist in the natural development of each human individual, helping us achieve complete human realization, mature development, of our human nature. A civilization that denies our subjectivity, including our transcendence as free persons seeking transcendent union with God, denies us what we are, our humanity, nature, completeness as persons. In the final analysis, such a misnamed civilization entombs us. It does not cultivate us.

We cannot treat science as if one kind of science exists. We cannot reduce rational knowledge to technology. The human race lives by reasoning in all cultural domains. We are especially interested in finding rational answers to our rational questions concerning ourselves and the purpose of our lives. As human beings, we have a natural moral and political right to seek rational answers to these questions.

When scientific leaders rejected the question "why" for the question of "know-how," they robbed us of the right to seek rational answers to the most rational of human questions. The single-minded quest for useful knowledge has helped technology's development. But it has also led our civilization into a blind alley. Scientists rejected the question "why" for irrational reasons, based upon *a priori* ideological assumptions.

Technology that does not respect the human subject or any higher end becomes a cruel instrument of destruction because, at its roots, it has separated itself from morality. By restoring an analogical conception of science, we can rationally employ the legitimately scientific question "why" to restore science's link with morality and the rest of culture and, once again, give culture an authentically human and rational face. We live by art and reasoning. We do not live only in the realm of technology. And technology is not science's only, or most important, domain.

When we no longer subordinate "know-how" to the question "why," then the question "know-how" starts to become meaningless. We can make sense of the question of "know-how" only if we know a thing's end. In science and culture as a whole we must restore the primary role of the question "why." Only then will we be able to seek knowledge completely rationally and use science, once again, to better, not destroy, culture in all its forms.

NOTES

Part One

Chapter One

1. Herodotus of Harlicarnassus, *The Histories*, no translator given (Harmonds-worth, Midx., England, Penguin Books, 1954), 2.49–2.50.
2. Plato *Timaeus*, English trans. R. G. Bury (Cambridge, Mass.: Harvard University Press, 1961); *Timajos* [*Timaeus*], Polish trans. Paweł Siwek (Warsaw: Państwowe Wydawnictwo Naukowe [State Scientific Publisher], 1986), 22A.
3. Aristotle, *Metaphysics*, English trans. Hugh Tredennick (Cambridge Mass.: Harvard University Press, Loeb Classical Library, 1933), *Metafizyka*, Polish trans. Tadeusz Żeleźnik, ed. Mieczysław Albert Krąpiec, Andrzej Maryniarczyk (Lublin: Redakcja Wydawnictw Naukowych [Editorial Board of Publishers], Catholic University of Lublin, 1996), Bk 1, ch. 1, 981b 23; Plato, *Phaedrus*, in *Euthyphro, Apology, Crito, Phaedo, Phaedrus,* English trans. Harold North Fowler (Cambridge, Mass.: Harvard University Press, Loeb Classical Library, 1914), 274C.
4. Charles Werner, *La philosophie grecque* [*Greek Philosophy*] (Paris: Payot, 1972), p. 9.
5. Arystoteles (Aristotle), *O niebie* [*On the Heavens*], Polish trans. Paweł Siwek, (Warsaw: Państwowe Wydawnictwo Naukowe, 1980); English trans. W. K. C. Guthrie, (Cambridge, Mass.: Harvard University Press, Loeb Classical Library, 1960), Bk. 2, ch. 12, 292a.
6. John Burnet, *Early Greek Philosophy* (New York: Meridian Books, 1930), pp. 1–30.
7. Ibid. pp. 18–19.
8. Plato, *The Laws*, English trans. R. G. Bury (Cambridge, Mass.: Harvard University Press, Loeb Classical Library, 1961); Prawa [*Laws, The*], Polish trans. Władysław Witwicki (Warsaw: Państwowe Wydawnictwo Naukowe, 1958), 819B–D.
9. Ibid. 819D; Marcus Tullius Cicero, *De republica*, in *De republica, De legibus*, English trans. Clinton Walker Keyes (Cambridge Mass.: Harvard University Press, repr. 1928 Loeb Classical Library ed., 1988), 1.29.
10. Burnet, *Early Greek Philosophy*, p. 19.
11. Herodotus, *The Histories*, 2.109.
12. Ibid.
13. Burnet, *Early Greek Philosophy*, pp. 19–21.
14. Ibid., pp. 21–22.
15. Ernan McMullin, *The Goals of Natural Science*, in "Proceedings and Addresses of the American Philosophical Association," 58:1 (1984), p. 38.
16. Ibid. p. 39.
17. Diogenes Laertius, *Lives of the Eminent Philosophers*, English trans. Robert Drew Hicks (Cambridge, Mass.: Harvard University Press, 1958–1959), Prol. , 1.3.

Chapter Two

1. [Arystoteles] (Aristotle), *Polityka* [*Politics*], Polish trans. Lodovici Piotrowicz (1964); English trans. Harris Rackham (Cambridge, Mass.: Harvard University Press, Loeb Classical Library, 1932), Bk. 7, 1327b.
2. See Werner Jaeger, *The Theology of the Early Greek Philosophers: The Gifford Lectures 1936* (London, Oxford, New York: Oxford University Press, 1968), pp. 9–10.
3. Ibid., p. 11.
4. Ibid., p. 10–12.
5. Ibid., p. 14.

Chapter Three

1. Werner Jaeger, *Paideia*, trans. from German to Polish Marian Plezia (Warsaw: Pax, 1962), vol. 1, p. 179; Plato, *Republic*, in *Platonis opera*, 5 vols., ed. John Burnet (Oxford: The Clarendon Press, 1900–1907; repr. 1950), Bk 9, 582C.
2. Aristotle, *Protrepticus*, trans. is based on a Polish trans. from the Greek by Kazimierz Leśniak, *Zachęta do filozofii* [*Exhortation to Philosophy*], (Warsaw: Państwowe Wydawnictwo Naukowe[State Scientific Publisher], 1988).
3. Ibid., fragment 20.
4. Ibid., fragment 20.
5. Ibid., fragment 21.
6. Ibid., fragment 23, 24.
7. Ibid., fragments 22–27.
8. Ibid., fragment 42.
9. Ibid., fragment 28.
10. Ibid., fragment 110.
11. See Plato, *Republic*.
12. Aristotle, *Protrepticus*, fragment 42.

Chapter Four

1. Cf. Werner Jaeger, *Paideia*, trans., from German to Polish Marian Plezia (Warsaw: Pax, 1962), vol. 1, p. 176ff.
2. Aristotle, *Metaphysics*, English trans. Hugh Tredennick (Cambridge Mass.: Harvard University Press, Loeb Classical Library, 1933), *Metafizyka*, Polish trans. Tadeusz Żeleźnik, ed. Mieczysław Albert Krąpiec, Andrzej Maryniarczyk (Lublin: Redakcja Wydawnictw Naukowych [Editorial Board of Publishers], Catholic University of Lublin, 1996), Bk. 1, 980b 25–28.
3. Ibid., 981a 5–12.
4. Ibid., Bks. 1–2. Cf. Mieczysław A. Krąpiec, *Filozofia—co wyjaśnia?* [*Philosophy— What does it explain?*] (Lublin: Lubelska Szkoła Filozofii Chrześcijańskiej [Lublin School of Christian Philosophy], 1999).
5. Aristotle, *Metaphysics*, English trans. Hugh Tredennick (Cambridge Mass.: Harvard University Press, Loeb Classical Library, 1933); *Metafizyka*, Polish trans. Tadeusz Żeleźnik, ed. Mieczysław Albert Krąpiec, Andrzej Maryniarczyk (Lublin: Redakcja Wydawnictw Naukowych [Editorial Board of Publishers], Catholic University of Lublin, 1996), Bk. 1981b 10–13.

6. Cf. Werner Jaeger, *The Theology of the Early Greek Philosophers* (London, Oxford, New York: Oxford University Press: 1968), ch. 2.

7. Aristotle, *Metaphysics*, Bk. 1, ch. 4 1070b 22.

8. Ibid., Bk. 1, ch. 12 982 b 12–20.

9. Ibid., 19–21.

10. Cf. Pierre Aubenque, *Le problème de l'être chez Aristotle* [*The Problem of Being in Aristotle*] (Paris: Presses Universitaires de France, 1977), pp. 28–44. Anton-Herman Chroust, "The Origin of Metaphysics," in *The Review of Metaphysics*, 14 (1960), pp. 601–617; Philippe Merlan, "On the Terms 'Metaphysics' and 'Being-qua-Being'" in *The Monist*, 52 (1968), pp. 174–194; Piotr Jaroszyński, "Spór o przedmiot 'Metafizyki' Arystotelesa" ["Dispute about the Object of Aristotle's Metaphysics"], in *Roczniki Filozoficzne* [*PhilosophicalYearbooks*], ed. Towarzystwo Naukowe, 31(1983), p. 1.

Chapter Five

1. *Słownik grecko-polski* [*Greek-Polish Dictionary*] (Warsaw: 1960), vol. 2, p. 514 ff.

2. Louis-Marie Régis, *L'Opinion selon Aristote* [*Opinion According to Aristotle*] (Paris, Ottawa: Institut d'études médiévales, 1935), p. 12.

3. Plato, *Republic*, in *Platonis opera*, 5 vols., ed. John Burnet (Oxford: The Clarendon Press, 1900–1907; repr. 1950), 511 A–E.

4. Ibid., 478 C–D.

5. Plato, *Theatetus*, 190 A.

6. Cf. Régis, *L'Opinion selon Aristote*, p. 50ff.

7. Plato, *Meno*, in *Platonis opera*, 97 B–C.

8. Plato, *Statesman*, in *Platonis opera*, 309 C.

9. Aristotle, *Posterior Analytics*, English trans. Hugh Tredennick, in Hugh Tredennick and E. S. Forster trans., *Posterior Analytics. Topica* (Cambridge Mass.: Harvard University Press, Loeb Classical Library,1934), Bk 1, ch. 4.

10. Aristotle, *Nicomachean Ethics*, trans. Harris Rackham (Cambridge, Mass.: Harvard University Press, Loeb Classical Library, 1934), Bk. 6, ch. 3, 1139 b 15–18.

11. Aristotle, *Topica*, English trans. E. S. Forster, in Hugh Tredennick and E. S. Forster trans. *Posterior Analytics. Topica*, Bk. 1, ch. 1; cf., also, *Nicomachean Ethics*, trans. Harris Rackham (Cambridge, Mass.: Harvard University Press, Loeb Classical Library, 1935), Bk. 3, ch. 22, 1111b 30.

12. Aristotle, *Posterior Analytics*, Bk. 1, ch. 33.

13. Ibid.

Chapter Six

1. Plato, *Phaedo*, in *Euthyphro, Apology, Crito, Phaedo, Phaedrus*, English trans. Harold North Fowler (Cambridge, Mass.: Harvard University Press, Loeb Classical Library, 1914), 96 B.

2. Ibid., 437 A.

3. Aristotle, *Physics*, English trans. Philip H. Wicksteed and Francis M. Cornford, (Cambridge, Mass.: Harvard University Press, Loeb Classical Library, 1960–1963), Bk. 7, ch. 3, 248a. For more on this, cf. Joseph Owens, *The Doctrine of Being in the*

Aristotelian "Metaphysics": *A Study in the Greek Background of Mediaeval Thought* (Toronto: Pontifical Institute of Mediaeval Studies, 1978), pp. 137–154.
4.   Ibid. p. 180–188.
5.   Aristotle, *Metaphysics*, English trans. Hugh Tredennick (Cambridge Mass.: Harvard University Press, Loeb Classical Library, 1933), *Metafizyka*, Polish trans. Tadeusz Żeleźnik, ed. Mieczysław Albert Krąpiec, Andrzej Maryniarczyk (Lublin: Redakcja Wydawnictw Naukowych [Editorial Board of Publishers], Catholic University of Lublin, 1996), Bk. 5, ch. 5, 1015 b 34–35.
6.   Mieczysław A. Krąpiec, Stanisław Kamiński, *Z teorii i metodologii metafizyki* [*On the Theory and Methodology of Metaphysics*] (Lublin: Redakcja Wydawnictw [Editorial Board of Publishers], Catholic University of Lublin, 1994), p. 220.
7.   Aristotle, *Posterior Analytics*, trans. Hugh Tredennick, in Hugh Tredennick and E. S. Forster trans., *Posterior Analytics. Topica* (Cambridge Mass.: Harvard University Press, Loeb Classical Library, 1934), Bk. 1, ch. 2; cf. Krąpiec, Kamiński, *Z teorii i metodologii metafizyki*, p. 221 ff.
8.   Krąpiec, Kamiński, *Z teorii i metodologii metafizyki*, p. 382 ff.
9.   Aristotle, *Posterior Analytics*, Bk. 2, ch. 19.
10.   Ibid.
11.   Aristotle, *On the Soul*, Bk. 2, chs. 5, 6, 7; Bk. 3, chs. 4, 5, 6.
12.   Owens, *The Doctrine of Being in the Aristotelian Metaphysics*, p. 431.

Chapter Seven

1.   Aristotle, *Metaphysics*, English trans. Hugh Tredennick (Cambridge Mass.: Harvard University Press, Loeb Classical Library, 1933), *Metafizyka*, Polish trans. Tadeusz Żeleźnik, ed. Mieczysław Albert Krąpiec, Andrzej Maryniarczyk (Lublin: Redakcja Wydawnictw Naukowych [Editorial Board of Publishers], Catholic University of Lublin, 1996), Bk. 3, ch. 4, 1000a 9–22.
2.   Mieczysław A. Krąpiec, Stanisław Kamiński, *Z teorii i metodologii metafizyki* [*The Theory and Methodology of Metaphysics*] (Lublin: Redakcja Wydawnictw [Editorial Board of Publishers] Catholic University of Lublin, 1994), p. 7.
3.   Aristotle, *Posterior Analytics*, trans. Hugh Tredennick, in *Posterior Analytics*, *Topica* (Cambridge Mass.: Harvard University Press, 1934), Bk. 1, ch. 2.
4.   Ibid.
5.   Ibid., ch. 2.
6.   Cf. Aleksander Achmanow, *Logika Arystotelesa* [*Aristotle's Logic*], trans. into Polish from the 1960 Russian edition A. Zabludowski and Barbara Stanosz (Warsaw: Państwowe Wydawnictwo Naukowe[State Scientific Publisher], 1965), p. 212
7.   Ibid., p. 224; Aristotle, *Prior Analytics*, Bk. 1, ch. 23.
8.   Aristotle, *Posterior Analytics*, Bk 2. ch. 2.
9.   Ibid., ch. 11.
10.   Aristotle, *Physics*, English trans. Philip H. Wicksteed and Francis M. Cornford, (Cambridge, Mass.: Harvard University Press, Loeb Classical Library,1960–1963), Bk 2, ch. 3, 195a 18–19; cf. *Introduction to Aristotle*, ed. Richard McKeon (New York: 1947), p. 88, n. 16.
11.   Aleksander Achmanow, *Logika Arystotelesa*, p. 221f f.
12.   Aristotle, *Posterior Analytics*, 71 b; cf. Achmanow, *Logika Arystotelesa*, p. 328 ff.

13. Aristotle, *Posterior Analystics*, Bk.2, ch. 7.

Chapter Eight

1. Plato, *Republic*, in *Platonis opera*, 5 vols., ed. John Burnet (Oxford: The Clarendon Press, 1900–1907; repr. 1950), 475 E.
2. Plato, *Listy* [*Letters*], trans. into Polish Maria Maykowska (Warsaw: Państwowe Wydawnictwo Naukowe [State Scientific Publisher]: 1987), Bk. 4, 320 B–C.
3. Plato, *Parmenides*, in *Platonis opera*, 136 D.
4. Plato, *Republic*, Bk. 9, 585 B–C.
5. Plato, *Phaedrus* [*Fajdros*], Polish trans. Leopold Regner (Warsaw: Państwowe Wydawnictwo Naukowe, 1993), 247 C–D.
6. Plato, *Philebus*, 64 E.
7. Plato, *The Laws*, English trans. R. G. Bury (Cambridge, Mass.: Harvard University Press, 1961), *The*[*Laws*], Polish trans. Władysław Witwicki (Warsaw: Państwowe Wydawnictwo Naukowe, 1958), Bk. 5, 730 B–C.
8. Plato, *Republic*, 489 E–450 A.
9. Plato, *Phaedo* [*Fedon*], Polish trans. Władysław Witwicki (Warsaw: Państwowe Wydawnictwo Naukowe, 1982), 65E–66C.
10. Plato, *Republic*, Bk. 3, 413 A.
11. Aristotle, *Metaphysics*, English trans. Hugh Tredennick (Cambridge Mass.: Harvard University Press, Loeb Classical Library, 1933), *Metafizyka*, Polish trans. Tadeusz Żeleźnik, ed. Mieczysław Albert Krąpiec, Andrzej Maryniarczyk (Lublin: Redakcja Wydawnictw Naukowych [Editorial Board of Publishers], Catholic University of Lublin, 1996), Bk. 5, ch. 7, 1017a 31–35.
12. Ibid., Bk. 9, ch, 9, 1051b 1–9.
13. Ibid., 1051b 13–18.
14. Ibid., 1051b 18–1052 a 5.
15. St. Thomas Aquinas, *Commentary on the Metaphysics of Aristotle*, trans. J. P. Rowan (Chicago: Henry Regnery Co., 1961,), Bk. 9, l. 11, nn. 1899, 1915.
16. Cf. Aleksander Achmanow, *Logika Arystotelesa*, ch. 5.
17. Cf. Aristotle, *On Interpretation*, Bk. 1, ch. 4.

Chapter Nine

1. Diogenes Laertius, *Lives of the Eminent Philosophers*, English trans. Robert Drew Hicks (Cambridge, Mass.: Harvard University Press, 1958–1959), 3. 84.
2. Aristotle, *Metaphysics*, English trans. Hugh Tredennick (Cambridge Mass.: Harvard University Press, Loeb Classical Library, 1933), *Metafizyka*, Polish trans. Tadeusz Żeleźnik, ed. Mieczysław Albert Krąpiec, Andrzej Maryniarczyk (Lublin: Redakcja Wydawnictw Naukowych [Editorial Board of Publishers], Catholic University of Lublin, 1996, Bk 6, ch. 1, 1025 b 21, 1026 b 5; Bk.11, ch. 7, 1064a 10, 1064a 17; *Topics*, trans. E. S. Forster, in *Posterior Analytics. Topica* (Cambridge Mass.: Harvard University Press, Loeb Classical Library, 1934), Bk. 4, ch. 6, 145a 15; Bk. 8, ch. 1, 57a 10; *Nicomachean Ethics*, trans. Harris Rackham (Cambridge, Mass.: Harvard University Press, Loeb Classical Library, 1934), Bk. 6, ch. 1, 1139a 27.
3. Cf. William David Ross, *Aristotle's Metaphysics* (Oxford: The Clarendon Press, 1981), vol. 1, p. 353.

238 SCIENCE IN CULTURE

4. Quintilian, *Institutes of Oratory*, trans. Harold Edgeworth Butler (Cambridge Mass.: Harvard University Press, Loeb Classical Library, 1959–1963), Bk. 2, ch. 18, 1–2.

Part Two

Chapter Ten

1. Cf. Pierre Aubenque, *Le problème de l'être chez Aristotle* [*The Problem of Being in Aristotle*] (Paris: Presses Universitaires de France, 1977), p. 23.
2. Cf. Edwin Bevan, *Stoics and Sceptics* (New York: Amo Press, 1979), ch. 4.
3. Cf. Marcia L. Colish, *The Stoic Tradition from Antiquity to the Early Middle Ages*, (Leiden: 1985), p. 9; cf. Georges Rodier, *Études de philosophie grecque* [*Studies of Greek Philosophy*] (Paris: J. Vrin, 1969), p. 222ff.
4. Cf. Max Pohlenz, *Die Stoa. Geschichte einter geistigen Bewegung* (Göttingen: Vandenhoeck and Ruprecht, 1964), p. 159.
5. Cf. Benson Mates, *Stoic Logic* (Berkeley, Los Angeles, London: California Library Reprint Series, 1973), p. 5.
6. Diogenes Laertius, *Lives and Views of the Eminent Philosophers*, English trans. Robert Drew Hicks (Cambridge, Mass.: Harvard University Press, Loeb Classical Library, 1958–1959), 7.5.
7. Cf. Colish, *The Stoic Tradition from Antiquity to the Early Middle Ages*, pp. 9–20.
8. Lucius Annaeus Seneca, *Ad Lucilium epistolae morales* [*Moral Epistles to Lucilius*], English trans. Richard M. Gummere (Cambridge, Mass.: Harvard University Press, 1996), 89.4–5, 8.
9. Clement of Alexandria, *Paedagogos*, vol. 2, 2, par. 25, 3, in *Protrepticus und Paedagogus* [*Exhortation and the Tutor*], ed. Otto Stälin (Leipzig: J. C. Hinrichs, 1905–1909); Karlheinz Hülser, *Die Fragmente zur Dialektik der Stoiker* (Stuttgart-Bad Cannstatt: 1987), p. 8ff.
10. Julius Domański, *Metamorfozy pojęcia filozofii* [*Metamorphoses of the Concept of Philosophy*], Polish trans. Z. Mroczkowska, Mirosław Bujko (Warsaw: Instytut Filozofii I Socjologii), p. 10ff.
11. Diogenes Laertius, *Lives and Views of the Eminent Philosophers*, 8.39.
12. Ernst Cassirer, *The Logic of the Humanities* (New Haven and London: Yale University Press 1966), p. 3.
13. Diogenes Laertius, *Lives and Views of the Eminent Philosophers*, 7.40; cf. Rodier, *Études de philosophie grecque*, pp. 247, 286.
14. Sextus Empricus, *Adversus mathematicos*, 1.16, in *Sextus Empiricus*, trans. R. G. Bury (Cambridge, Mass.: Harvard University Press, Loeb Classical Library, 1939–1949), vol. 1.
15. Plato, *Statesman*, in *Platonis opera*, 5 vols., ed. John Burnet (Oxford: The Clarendon Press, 1900–1907; repr. 1950), 258 E.
16. Richard Heinze, *Xenokrates. Darstellung der Lehre und Sammlung der Fragmente* (Hildesheim: 1965), p. 1ff.
17. Diogenes Laertius, *Lives and Views of the Eminent Philosophers*, 7.1; cf. Rodier, *Études de philosophie grecque*, p. 302ff.

Notes

18. Marcus Tullius Cicero, *On the Highest Good and Evil*, 3.5, 16; Diogenes Laertius, *Lives and Views of the Eminent Philosophers*, 7.85; Seneca's quote is from, *Epistulae morales ad Lucilium*, 20.121.

19. Cf. G. Rodier, *Études de philosophie grecque,*, pp. 302–306.

20. Karlheinz Hülser, *Die Fragmente zur Dialektik der Stoiker*, vol. 1, p. 22; cf. Rodier, *Études de philosophie grecque*, p. 247.

21. Émile Bréhier, *Histoire de la philosophie* [*History of Philosophy*] (Paris: Libraire Félix Alcan, 1931), vol. 1, p. 299.

22. Pierre Hadot, *"Philosophie, discours philosophiques, et divisions de la philosophie chez les stoiciens"* ["Philosophy, Philosophical Discourses, and Stoic Divisions of Philosoophy"], in *Revue international de la philosophie*, 45:178 (March 1991), p. 209.

23. Philonis Alexandrini, *De mutatione nominum*, in *Opera quae supersunt*, Berlin 1896–1926, vol. 3. Cf. Karlheinz Hülser, *Die Fragmente zur Dialektik der Stoiker*, p. 18.

24. Cf. Rodier, *Études de philosophie grecque* , p. 290.

25. Marcia L. Colish, *The Stoic Tradition from Antiquity to the Early Middle Age*, p. 44.

26. Ibid.

27. Cf. Rodier, *Études de philosophie grecque* , p. 249.

28. St. Thomas Aquinas, *Quaestiones disputatae de veritate*, in *Sancti Thomae Aquinatis, doctoris angelici Opera Omnia iussu Leonis XIII. O.M. edita, cura et studio fratrum praedicatorum* (Rome: 1882–1996), 2, 6c.

29. E.g., Colish, *The Stoic Traditio from Antiquity to the Early Middle Ages*, p. 22–36; Michel Spannent, *Permanence du stoicisme. De Zenon á Malraux* [*The Permanence of Stoicism: From Zeno to Malraux*] (Gembloux 1973), pp. 23–25; Rodier, *Études de philosophie grecque*, p. 245)

30. Cf. Aristotle, *Metaphysics*, English trans. Hugh Tredennick (Cambridge Mass.: Harvard University Press, Loeb Classical Library, 1933), *Metafizyka*, Polish trans. Tadeusz Żeleźnik, ed. Mieczysław Albert Krąpiec, Andrzej Maryniarczyk (Lublin: Redakcja Wydawnictw Naukowych [Editorial Board of Publishers], Catholic University of Lublin,1996), Bk. 1, chs. 3–4.

31. Aristotle, *Physics*, English trans. Philip H. Wicksteed and Francis M. Cornford, (Cambridge, Mass.: Harvard University Press, 1960–1963), Bk. 1, ch. 2, 184b; ch. 9, 192a.

32. Bréhier, *Histoire de la philosophie* , vol. 1, 2, p. 297ff.

33. Cf. Hülser, *Die Fragmente zur Dialektik der Stoiker*, vol. 1, p. XVIII ff.

34. Émile Bréhier, *Chrysippe et l'ancien stoicisme* [*Chrysippus and Ancient Stoicism*] (Paris: Presses Universitaires de France, 1951).

35. Gerard Verbecke, *"La philosophie du signe chez les stoïciens"* ["The Philosophy of the Sign among the Stoics," in *Les stoiciens et leur logique* [*The Stoics and their Logic*], ed. J. Brunschwig (Paris: J. Vrin, 1978), p. 401.

36. Ibid., p. 402, 418n. 7.

37. Cf. Gerard Verbeke, *"Philosophie et sémiologie chez les stoïciens"* [*"Philosophy and Semiology in the Stoics"*], in *Études philosophiques présentées au Dr. Ibrahim Madkour* [*Philosophical Studies Presented to Dr. Ibrahim Madkour*], no editor given (Cairo, General Egyptian Book Organization, 1974), p. 15–38.

38. Gerard Verbeke, *"La philosophie du signe chez les stoïciens,"* p. 401.

240   *SCIENCE IN CULTURE*

Chapter Eleven

1. Cf. Richard T. Wallis, *Neoplatonism* (London: Duckworth, 1972), p. 1.
2. Ibid., pp. 37–47.
3. Ibid. p. 13.
4. Plotyn [Plotinus], *Ennedy* [*Enneads*], Polish trans. Adam Krokiewicz (Warsaw: Państwowe Wydawnictwo Naukowe[State Scientific Publisher],1959), Bk. 6, ch. 9, 11.
5. Cf. Thomas Whittaker, *The Neo-Platonists: A Study in the History of Hellenism* (Hildesheim, Zürich, New York: 1987), ch. 6.
6. Cf. Wallis, *Neoplatonism*, p. 13–15.
7. Cf. Gregory Shaw, *Theurgy and the Soul: The Neoplatonism of Iamblichus* (Pennsylvania: : The Pennsylvania State University Press, 1995), p. 2.
8. E. des Places (ed. and trans.), *Jamblique: Les mystères d'Egypte* (Paris: Les Belles Lettres, 1966), p. 259, 5–14; cf. Shaw, *Theurgy and the Soul: The Neoplatonism of Iamblichus*, pp. 3, 6, 7.
9. Plato, *The Laws*, English trans. R. G. Bury (Cambridge, Mass.: Harvard University Press, Loeb Classical Library, 1961); *Prawa* [*Laws, The*], Polish trans. Władysław Witwicki (Warsaw: Państwowe Wydawnictwo Naukowe, 1958), 657 A. On the influence of the deeper mysteries upon Plato's thought see his *Seventh Letter*, 341 C–D; on the influence of Egyptian "wisdom" on his views, see *Statesman* 290 C–E; *Timaeus*, 21 D–25 D; *Phaedrus*, 275B; *Laws*, 819 B; *Philebus*, 18B; *Charmides*, 156B–157C); cf. Shaw, *Theurgy and the Soul: The Neoplatonism of Iamblichus*, p. 6.
10. Des Places, *Jamblique: Les mystères d'Egypte*, p. 258, 3–6.
11. Ibid, p. 259, 14–19; cf. Shaw, *Theurgy and the Soul: The Neoplatonism of Iamblichus*, p. 3ff.
12. Wallis, *Neoplatonism*, p. 108.
13. Cf. H. Lewy, *Chaldean Oracles and Theurgy: Mysticism, Magic, and Platonism in the Later Roman Empire* (Paris: Études Augustiniennes, 1978).
14. Cf. Wallis, *Neoplatonism*, p. 105; cf., also G. Shaw, *Theurgy and the Soul: The Neoplatonism of Iamblichus*, p. 6.
15. Lewy, *Chaldean Oracles and Theurgy. Mysticism, Magic, and Platonism in the Later Roman Empire*, p. 461.
16. Ibid., p. 462; des Places, *Jamblique: Les mystères d'Egypte*, p. 2.
17. Plato, *Phaedrus*, 244 A ff.
18. Lewy, *Chaldean Oracles and Theurgy: Mysticism, Magic, and Platonism in the Later Roman Empire*, p. 463.
19. Wallis, *Neoplatonism*, p. 107.
20. Lewy, *Chaldean Oracles and Theurgy. Mysticism, Magic, and Platonism in the Later Roman Empire*, p. 487–489.
21. Wallis, *Neoplatonism*, p. 107.
22. Ibid.
23. Plotinus *Enneads*, Bk. 5, ch. 9, 2; cf. Plato, *Uczta* [*Symposium*], Polish trnas. Władysław Witwicki (Warsaw: Państwowe Wydawnictwo Naukowe, 1975), 210 A ff.
24. Wallis, *Neoplatonism*, p. 108.
25. Des Places, *Jamblique: Les mystères d'Egypte*, p. 2, 2; Wallis, *Neoplatonism*, p. 120.
26. Plotinus, *Enneads*, Bk. 3, 1.3, 13–17.
27. Des Places, *Jamblique: Les mystères d'Egypte*, p. 3, 3.

28. Ibid., 184 1–8; see also, Shaw, *Theurgy and the Soul: The Neoplatonism of Iamblichus*, p. 51.
29. Shaw, *Theurgy and the Soul: The Neoplatonism of Iamblichus*, p. 57.

Part Three

Chapter Twelve

1. Cf. "Tatian," in *The Catholic Encyclopedia* (New York: 1912), vol. 14, p. 464.
2. Cf. Hamilton Baird Timothy, *The Early Christian Apologists and Greek Philosophy*, (Assen: Van Gorcum, 1973), p. 40.
3. Ibid., p. 54 ff.
4. Ibid., p. 41–48.
5. Ibid., p. 58.
6. Étienne Gilson, *Études de philosophie médiévale* (Strasbourg: Commission des publications de la Faculté des lettres de l'Université de Strasbourg, 1921), p. 31.
7. Ibid., p. 37.
8. Ibid., p. 32ff.
9. Peter Damian, *De divina omnipotentia et altri opuscoli* [*Concerning Divine Omnipotence and Other Works*], ed. Paolo Brezzi, trans. Bruno Nardi (Firenze, Vallecchi: 1943). Cited after Gilson, *Études de Philosophie médiévale*, p. 36.
10. Ibid., pp. 44–46.

Chapter Thirteen

1. A. Tribbechovia, *De doctoribus scholasticis* [*Concerning the Scholastic Doctors*] (Jena: 1719), p. xxii. Cited after Bernardus Baudouw, "*Philosophia 'Ancilla Theologiae'*" ["Philosophy 'Handmaiden to Theology'"], in *Antonianum*, Annus 12 (Rome: 1937) p. 306n1.
2. Aristotle, *Metaphysics*, English trans. Hugh Tredennick (Cambridge Mass.: Harvard University Press, Loeb Classical Library, 1933), *Metafizyka*, Polish trans. Tadeusz Żeleźnik, ed. Mieczysław Albert Krąpiec, Andrzej Maryniarczyk (Lublin: Redakcja Wydawnictw Naukowych [Editorial Board of Publishers], Catholic University of Lublin, 1996), Bk 2, ch. 2.
3. Ibid., Bk. 6.
4. Bernardus Baudoux, "*Philosophia 'Ancilla Theologiae*," p. 295n4. On the role of Philo in the beginning of so-called Christian philosophy as the handmaiden of theology, see Peter A. Redpath, *Wisdom's Odyssey from Philosophy to Transcendental Sophistry* (Amsterdam and Atlanta: Value Inquiry Book Series [VIBS], Editions Rodopi, B. V., 1997), p. 33 ff.
5. Baudoux, "*Philosophia 'Ancilla Theologiae*," p. 296 ff.
6. Ibid., p. 299.
7. Ibid., p. 300 ff.
8. Ibid., p. 304.

242   *SCIENCE IN CULTURE*

Chapter Fourteen

1. Marie-Dominique Chenu, *La théologie comme science au XIIIe siècle*
[*Theolgy as Science in the 13th Century*] (Paris 1957), p. 15.
2. Ibid., p. 16 ff.
3. Ibid., p. 18, note 1, and ff.
4. Ibid., p. 20.
5. Ibid., p. 25.
6. Ibid., p. 20 f.
7. *Hebrews*. 11, 1.
8.   St. Aurelius Augustine, *De praedestinatione sanctorum* [*On the Predestination of the Blessed*], ii, in *Patrologia latina*, ed. Migne (Paris: 1978–1990), vol. 44.
9.   St. Augustine, *Epistola*, in *Patrologia latina*, ed. Migne (Paris: 1978–1990), vol. 44, 120, 3; cf. Charles Norris Cochrane, *Christianity and Classical Culture: A Study of Thought and Action from Augustus to Augustine* (Oxford: Oxford University Press, 1957).
10.   Cf. Chenu, *La théologie comme science au XIIIe siècle*, p. 34.
11.   Ibid., p. 37.
12.   Ibid., pp. 38–41.
13.   Ibid., p. 39.
14.   Ibid., p. 54.
15.   Ibid., p. 55.
16.   Ibid., p. 56.
17.   Ibid., p. 59.
18.   Ibid.
19.   Ibid.
20.   St. Thomas Aquinas, *In I Sententiarum, Scriptum super libros Sententiarum*, ed. Pierre Mandonnet (Paris: 1929), vol. 1, prol., a. 3, sol. 2, ad 2; Chenu, *La théologie comme science du XIIIe siècle*, p. 64.
21.   St. Thomas Aquinas, *In III Sententiarum, Scriptum super libros Sententiarum*, ed. M. F. Moos (Paris, 1933), vol. 3, d. 24, q. 1, a. 2, sol. 1, obj. 2 and ad 2; Chenu, *La théologie comme science du XIIIe siècle*, p. 65.
22.   St. Thomas Aquinas, *Summa theologiae*, in *Sancti Thomae Aquinatis, doctoris angelici Opera Omnia iussu Leonis XIII. O.M. edita, cura et studio fratrum praedicatorum* (Rome: 1882–1996), 1, q.1, a. 2, ad 2; Chenu, *La théologie comme science du XIIIe siècle*, p. 72.
23.   St. Thomas Aquinas, *Quaestiones disputatae de veritate*, in *Sancti Thomae Aquinatis, doctoris angelici Opera Omnia iussu Leonis XIII. O.M. edita, cura et studio fratrum praedicatorum* (Rome: 1882–1996), vol. 22, q.14, a. 9, ad 3; Chenu, *La théologie comme science du XIIIe siècle* , p. 73).
24.   St. Thomas Aquinas, *Expositio super librum Boethii de Trinitate*, in *Opusculae theologiae*, ed. Marietti (Turin, Rome: 1954), vol. 2, q. 2, a. 3, ad 5; Chenu, *La théologie comme science du XIIIe siècle*, p. 86.
25.   Chenu, *La théologie comme science du XIIIe siècle*, p. 78.

Chapter Fifteen

1. Cf. Joseph Owens, *The Doctrine of Being in the Aristotelian "Metaphysics"*: *A Study in the Greek Background of Mediaeval Thought* (Toronto: Pontifical Institute of Mediaeval Studies, 1978).
2. Aristotle, *Metaphysics*, English trans. Hugh Tredennick (Cambridge Mass.: Harvard University Press, Loeb Classical Library, 1933), *Metafizyka*, Polish trans. by Tadeusz Żeleźnik, ed. Mieczysław Albert Krąpiec, Andrzej Maryniarczyk (Lublin: Redakcja Wydawnictw Naukowych [Editorial Board of Publishers], Catholic University Lublin, 1966), Bk. 6, ch. 1.

Chapter Sixteen

1. Marie-Dominique Chenu, *La théologie comme science au XIIIe siècle* [*Theology as Science in the 13th Century* (Paris: J. Vrin 1957), p. 85.
2. St. Thomas Aquinas. *Expositio super librum Boethii de Trinitate*, in *Opusculae theologica*, ed. Marietti (Turin, Rome: 1954), vol. 2, q. 2, a.. 3.; Chenu, *La théologie comme science du XIIIe siècle*, p. 88.
3. St. Thomas Aquinas, *Expositio super Librum Boethii de Trinitate.*, q. 2, a. 3.
4. St. Thomas Aquinas, *Quaestiones disputatae de veritate*, in *Sancti Thomae Aquinatis, doctoris angelici Opera Omnia iussu Leonis XIII. O.M. edita, cura et studio fratrum praedicatorum* (Rome: 1882–1996), vol. 22, q. 14, a. 9, ad 3.
5. *"Rozum otwarty na wiarę. 'Fides et ratio'—w rocznicę ogłoszenia. 2 Międzynarodowe Sympozjum Metafizyczne, Katolicki Uniwersytet Lubelski* [*Catholic University of Lublin*], 9–10 December 1999" ["Reason Open to Faith —Second International Metaphysical Symposium on the Anniversary of the Publication of *'Fides et ratio*,' Lublin, Catholic University of Lublin, 9–10 December 1999"]; cf. Mieczysław A. Krąpiec, *Filozofia w teologii* [*Philosophy in Theology*] (Lublin, 1998), pp. 121, 125.
6. Pope John Paul II, *Fides et ratio*, ch. 44; see also chs. 43, 62, 66, 69, 77, 78, 79.

Part Four

Chapter Seventeen

1. Herbert Butterfield, *The Origins of Modern Science 1300–1800* (London: G. Bell and Sons, Ltd., 1949), p. 163.
2. David C. Lindberg, *The Beginnings of Western Science* (Chicago and London: University of Chicago Press, 1992), pp. 147–149.
3. Alistair Cameron Crombie, *Nauka średniowieczna i początki nauki nowożytnej* [*Mediaeval Science and the Beginnings of Modern Science*], trans. into Polish S. Łypacewicz (Warsaw 1960), vol. 1, p. 27.
4. Lindberg, *The Beginnings of Western Science*, p. 156.
5. Ibid., p. 158.
6. Cf. Norman H. Baynes and L. B. Moss, *Byzantium. An Introduction to East Roman Civilization* (Oxford: The Clarendon Press, 1949), p. 201 ff.
7. Ibid., p. 206.

8. Lindberg, *The Beginnings of Western Science*, p. 161 ff.
9. Ibid., p. 163.
10. Nina W. Pigulewska, *Katolicki Uniwersytet Lubelski* [*Catholic University of Lublin*]. *Kultura syryjska we wczesnym średniowieczu* [*Syrian Culture in the Early Middle Ages*] (Warsaw: Pax, 1989), p. 121 ff.
11. Ibid., p. 129.
12. Ibid., p. 161.
13. Lindberg, *The Beginnings of Western Science*, pp. 168–170.
14. Ibid., p. 180–181. Cf. John N. Deely, *Four Ages of Understanding. The First Postmodern Survey of Philosophy from Ancient Times to the Turn of the Twenty-First Century* (Toronto, Buffalo, London: University of Toronto Press, 2001), pp. 186–193.
15. Ibid., pp. 204–206.
16. Crombie, *Nauka średniowieczna i początki nauki nowożytnej*, vol. 1, ch. 2.
17. Butterfield, *The Origins of Modern Science 1300–1800*, p. 67.

Chapter Eighteen

1. William Whewell, *History of the Inductive Sciences from the Earliest Times to the Present Time* (New York: Dover, 1959), vol. 1, p. 245; cf. Jeremiah Hackett, "Roger Bacon on *Scientia Experimentalis*," in *Roger Bacon and the Sciences*, ed. Jeremiah Hackett (Leiden, New York, Köln: E. J. Brill, 1997), p. 279.
2. Ibid., pp. 280–283.
3. Stewart C. Easton, *Roger Bacon and his Search for a Universal Science* (Westport, Conn.: Greenwood Press, 1970), p. 77 ff.
4. Alistair Cameron Crombie, *Nauka średniowieczna i początki nauki nowożytnej* [*Mediaeval Science and the Beginnings of Modern Science*], Polish trans. S. Łypacewicz (Warsaw: Pax, 1960), vol. 1, p. 72.
5. Easton, *Roger Bacon Roger Bacon and his Search for a Universal Science*, p. 97.
6. Roger Bacon, *Opus tertium*, 12, cited after Easton, *Roger Bacon and his Search for a Universal Science*, p. 74.
7. Plato, *Gorgias*, in *Plato: Lysis, Symposium, Gorgias*, Walter R. M. Lamb trans. (Cambridge Mass.: Harvard University Press, Loeb Classical Library, 1961), 448 C.
8. Plato, *Timaeus*, in *Timaeus, Critias, Cleitophon, Menexenus, Letters*, R. G. Bury trans. (Cambridge Mass.: Harvard University Press, 1961), 68 D.
9. Arystoteles (Aristotle), *Posterior Analytics*, in *Analityki pierwsze i wtóre* [*Prior and Posterior Analytics*], trans. Kazimierz Leśniak (Warsaw: Państwowe Wydawnictwo Naukowe [State Scientific Publisher], 1973), Bk. 2, ch. 19, 100a.
10. Aristotle, *Metaphysics*, English trans. Hugh Tredennick (Cambridge Mass.: Harvard University Press, Loeb Classical Library, 1933), *Metafizyka*, Polish trans. Tadeusz Żeleźnik, ed. Mieczysław Albert Krąpiec, Andrzej Maryniarczyk (Lublin: Redakcja Wydawnictw Naukowych [Editorial Board of Publishers], Catholic University of Lublin, 1996), Bk. 1, ch. 1, 980 b 29–981 a 17.
11. Cf. Roman Bugaj, *Hermetyzm* [*Hermeticism*], (Wrocław, Warsaw, Kraków: Ossolineum, 1991), p. 103ff.
12. Cf. Alistair Cameron Crombie, *Robert Grosseteste and the Origins of Experimental Science 1100–1700* (Oxford: The Clarendon Press, 1953), p. 16f f.
13. Hackett, "Roger Bacon on *Scientia Experimentalis*," p. 289ff.

14. Ibid., p. 290ff.
15. Alistair Cameron Crombie, *Style myśli naukowej w początkach nowożytnej Europy* [*Styles of Scientific Thought in the Beginnings of Modern Europe*], trans. into Polish Piotr Salwa (Warszawa: Polska Akademia Nauk, Instytut Filozofii i Socjologii 1994), p. 35ff
16. William Eamon, "From the Secrets of Nature to Public Knowledge," in *Reappraisals of the Scientific Revolution*, ed. David C. Lindberg, Robert S. Westman (Cambridge, England: Cambridge University Press, 1990), p. 337.
17. Cf. Lyon Sprague de Camp, *Wielcy i mali twórcy cywilizacji* [*Major and Minor Creators of Civilization*], Polish trans. B. Orłowski (Warsaw: Wiedza Powszechna), 1968, p. 121ff.
18. Ibid., p. 124–126.
19. Ibid., pp. 149–156.
20. Ibid., p. 173f f.
21. Ibid., pp. 188–203.
22. Roger Bacon, *Epistola de secretis operibus artis et naturae*, cited after Friedrich Klemm, *A History of Western Technology* (Cambridge, Mass.: Harvard University Press, 1964), p. 95.

Chapter Nineteen

1. Plato, *Timaeus*, in *Timaeus, Critias, Cleitophon, Menexenus, Letters*, English trans. R. G. Bury (Cambridge Mass.: Harvard University Press, Loeb Classical Library, 1961), 31 C.
2. Ibid., 53 C–D.
3. Ibid., 54 B–C.
4. Cf. Alistair Cameron Crombie, *Robert Grosseteste and the Origins of Experimental Science 1100–1700* (Oxford: The Clarendon Press,1953), p. 4 ff.
5. Cf. Philippe Merlan, *From Platonism to Neoplatonism* (The Hague: Martinus Nijhoff, 1960).
6. Alistair Cameron Crombie, *Robert Grosseteste and the Origins of Experimental Science 1100–1700*, p. 104.
7. Ibid., p. 108ff.
8. Ibid., p. 11.
9. Ibid., p. 134.

Chapter Twenty

1. Joseph Owens, *The Doctrine of Being in the Aristotelian "Metaphysics": A Study in the Greek Background of Mediaeval Thought* (Toronto: Pontifical Institute of Mediaeval Studies, 1978), p.108 ff.
2. Porphyry, *Isagoge*, ch. 1, in Aristotyles (Aristotle*), Kategorie i Hermeneutyka z dodaniem Isagogi Porfiriusza*, [*Categories and Hermeneutics with the addition of Porphyry's Isagoge*], trans. into Polish by Kazimeirz Leśniak (Warsaw: Państwowe Wydawnictwo Naukowe [State Scientific Publisher], 1975).
3. Cf. Mieczysław A. Krąpiec, *Byt i istota* [*Being and Essence*] (Lublin: Redacja Wydawnictw [Editorial Board of Publishers], Catholic University of Lublin, 1994), chs. 5, 6.

4. Jean Largeault, *Enquête sur le nominalisme* [*Inquiry on Nominalism*] (Paris, Béatrice-Nauwelaerts; Louvain, Éditions Nauwelaerts: 1971), p. 12.
5. Cf. Mieczysław A. Krąpiec, *Metaphysics: An Outline of the History of Being*, Theresa Sandok trans. (New York: Peter Lang 1991), pp. 407–443.
6. Ibid.
7. Ibid.

Part Five

Chapter Twenty-One

1. Charles Webster, *The Great Instauration: Science, Medicine, and Reform 1626–1660* (London: Duckworth, 1975), p. 347.
2. Francis Bacon, *Novum organum* (New York: Collier, 1901); Polish trans. Jan Wikarjak (Warsaw: Państwowe Wydawnictwo Naukowe [State Scientific Publisher], 1955), 2.52.
3. *Daniel*, 12, 1–4.
4. Webster, *The Great Instauration: Science, Medicine, and Reform 1626–1660*, p. 2.
5. John Milton, *Prolusion*, 7, CPW, i, p. 296; cited after Webster, *The Great Instauration: Science, Medicine and Reform 1626–1660* , p. 1.
6. Webster, *Science, Medicine, and Reform 1626–1660*, p. 16.
7. Ibid., p. 14.
8. Ibid.
9. Ibid., pp. 326–327.
10. Ibid., p. 18.
11. Cf. P. M. Rattansi, "The Social Interpretation of Science in the Seventeenth Century, " in *Science and Society 1600–1900*, ed. Peter Mathias (Cambridge, England: Cambridge University Press, 1972), p. 12.
12. Webster, *The Great Instauration: Science, Medicine, and Reform 1626–1660*, p. 16–17.
13. Francis Bacon, *Novum organum* (New York: Collier, 1901), Polish trans. Jan Wikarjak (Warsaw: Państwowe Wydawnictwo Naukowe, 1955), 1. 92. Compare Pado Rossi, "Bacon's Idea of Science," in *The Cambridge Companion to Bacon*, ed. Markku Peltonen (Cambridge, England: 1996), p. 26.
14. Herbert Butterfield, *The Origins of Modern Science 1300–1800* (London: G. Bell and Sons, Ltd., 1949), pp. 165, 166, 174.

Chapter Twenty-Two

1. Arystoteles (Aristotle), *Polityka* [*Politics*], Polish trans. Lodovici Piotrowicz (1964); English trans. Harris Rackham (Cambridge, Mass.: Harvard University Press, 1932), Bk. 8, Ch. 2.
2. Marcus Tullius Cicero, *De re publica*, in *De republica, De legibus*, English trans. Clinton Walker Keyes (Cambridge Mass.: Harvard University Press, Loeb Classical Library, 1988), 3.6, 27.
3. St. Aurelius Augustine, *"Sermo* [*Sermon*], 37, 5, 6, " *New Advent*, URL= http://www.newadvent.org/fathers/160337.htm; Pope John Paul II, *Laborem exercens*

(Wrocław: Wrocławskie Wydawnictwo Oświatowe, 1992); *On Human Work: Laborem exercens* (Washington, D. C.: Distributed for the United States Catholic Conference, Hunter Publishing, 1981), passim.
   4.   Cf. Pado Rossi, *Francis Bacon: From Magic to Science* (Chicago: University of Chicago Press, 1978), p. 1.
   5.   Ibid., p. 2–7.
   6.   Ibid., p. 8.
   7.   Jacques Maritain, *Three Reformers: Luther—Descartes—Rousseau* (London: Sheed & Ward, 1950), p. 32.
   8.   Ibid., p. 33.
   9.   Ibid.
   10.   Francis Bacon, *Novum organum* (New York: Collier, 1901), *Novum organum*, Polish trans. Jan Wikarjak (Warsaw: Państwowe Wydawnictwo Naukowe [State Scientific Publisher], 1955), p. 23; *The Great Instauration*, in *The English Philosophers from Bacon to Mill*, ed. E. A. Burtt (New York: 1939), p. 13.
   11.   Bacon, *Novum organum*, p. 28; *The Great Instauration*, p. 15.
   12.   Bacon, *Novum organum*, p. 105.
   13.   Ibid., p. 124. A speech by Prince Albert in 1850 at a Banquet at the Mansion House, cited after: *The Open University: An Arts Foundation Course A 102 Units 31–32 Conclusion* (Walton Hall, Milton Keynes: 1987), p. 30.

Chapter Twenty-Three

   1.   Cf. Roman Bugaj, *Hermetyzm* [*Hermeticism*] (Wrocław, Warsaw, Kraków: Ossolineum, 1991), p. 166.
   2.   Cf. Daniel Pickering Walker, *The Ancient Theology: Studies in Christian Platonism from the Fifteenth to the Eighteenth Century* (London: Duckworth, 1972), p. 17.
   3.   Ibid., p. 20ff.; Pado Rossi, *Francis Bacon: From Magic to Science* (Chicago: University of Chicago Press, 1978), pp. 69–70.
   4.   Ibid., p. 18ff.
   5.   Martin Bernal, *Black Athena: The Afroasiatic Roots of Classical Civilization* (London: Free Association Books, 1987), pp. 161–163, 165.
   6.   Ibid., p. 17.
   7.   For example, see Francis A. Yates, *The Occult Philosophy in the Elizabethan Age* (London, Boston: Routledge & Kegan Paul, 1979); *Giordano Bruno et la tradition hermetique* [*Giordano Bruno and the Hermetic Tradition*] (Paris: Dervy-Livres, 1988).
   8.   André-Jean Festugière, *La Révélation d'Hermes Trismegiste* [*The Revelation of Hermes Trismegistus*] (Paris: Lecoffre, 1950–1954), vols. 1–4.
   9.   Cf. P. M. Rattansi, "The Social Interpretation of Science in the Seventeenth Century," in *Science and Society 1600–1900*, ed. Peter Mathias (Cambridge, England: Cambridge University Press, 1972), p. 9.
   10.   Cf. Garth Fowden, *The Egyptian Hermes: A Historical Approach to the Late Pagan Mind* (Princeton: Princeton University Press, 1986), p. 22 ff.
   11.   Ibid., p. 89 ff.
   12.   Hermann Diels, *Die fragmente der Vorsokratiker, griechisch und deutsch* [*The Fragments of the Presocratics: Greek and German*]. (Berlin: Weidmannsche Buchhandlung, 1906–1910), B112. 4–5. Peter Kingsley, *Ancient Philosophy, Mystery, and Magic:*

Empedocles and Pythagorean Tradition (Oxford: The Clarendon Press, 1995), pp. 229, 271–391

13. Diels, *Die fragmente der Vorsokratiker, griechisch und deutsch*, B112–114; Kingsley, *Ancient Philosophy, Mystery, and Magic: Empedocles and Pythagorean Tradition*, pp. 229, 271–391; Hippocrates, *De morbo sacro*, in Hippocrates, W*orks*, W. H. S. Jones trans. (Cambridge, Mass.: Harvard University Press, Loeb Classical Library, 1959–1962).

14. Fowden, *The Egyptian Hermes: A Historical Approach to the Late Pagan Mind*, p. 275.

15. Brian P. Copenhaver, "Hermes Trismegistus and Early Modern Science: The 'Yates Thesis,'" in *Reappraisals of the Scientific Revolution*, ed. David C. Lindberg, Robert S. Westman (Cambridge, Mass.: Harvard University Press, 1990), p. 266 ff.

16. Ibid., pp. 272, 289.

17. Ibid., p. 275 ff.

18. Ibid., p. 280 ff.

19. Cornelius Agrippa, *De incertitudine et vanitate scientiarum* (Kolonia: 1527); cited after Rossi, *Francis Bacon: From Magic to Science*, p. 18–19.

20. Rossi, "Bacon's Idea of Science," *The Cambridge Companion to Bacon*, ed. Markku Peltonen (Cambridge, England: 1996), pp. 30–31

21. Ibid., p. 91 ff.

22. Cited after P. Rossi, *Francis Bacon: From Magic to Science*, p. 29.

23. Cf. William Eamon, "From the Secrets," in *Reappraisals of the Scientific Revolution*, pp. 334–337.

24. Eamon, *From the Secrets*, p. 355.

25. Rossi, "Bacon's Idea of Science," in *The Cambridge Companion to Bacon*, pp. 32, 34.

26. Cf. Francis A. Yates, *Giordano Bruno et la tradition hermétique*, p. 177.

27. Cf. P. M. Rattansi, "The Social Interpretation of Science in the Seventeenth Century," p. 16.

28. Giovanni Pico della Mirandola, *Oration on the Dignity of Man*, trans. A. Robert Caponigri (Chicago: Regnery Publishing, 1956); cited after URL=http://en.wikipedia.org/wiki/Giovanni_Pico_della_Mirandola.

29. Ibid.

30. Rattansi, *The Social Interpretation of Science in the Seventeenth Century*, in *Science and Society 1600–1900*, p. 5.

31. Ibid., p. 6, 16.

32. "Cabala," in *The Jewish Encyclopedia*, no editor given (New York and London: 1902), vol. 3, p. 456 ff.

33. Francis A. Yates, *The Occult Philosophy in the Elizabethan Age* (London, Boston: Henley, 1979), pp. 1–9.

34. Ibid., p. 12.

35. *The Jewish Encyclopedia*, p. 40.

36. Cf. Yates, *The Occult Philosophy in the Elizabethan Age*, pp. 10–15.

37. Étienne Gilson, *History of Christian Philosophy in the Middle Ages* (London: 1955), p. 352).

38. Yates, *The Occult Philosophy in the Elizabethan Age*, p. 20.

39. Ibid., p. 25.

40. Cf. Anne Becco, "Leibniz et François-Mercure von Helmont: Bagatelle pour des monades," in *Magia naturalis und die Entstehung der modernen Naturwissenschaften, Symposion der Leibniz-Gesellschaft [Magia naturalis and the Emergence of the Natural Sciences, Symposium of the Leibniz Society]*, Hannover, 14–15 November 1975, ed. Albert Heinekamp and Dieter Mettler (Wiesbaden: Franz Steiner Verlag, 1978), pp. 119–141. See also, *Cabbala Denudata*, tom. 1, part. 2, p. 310, sqq, et tom. 2, de tract. ult. p. 28, 4.5.

41. "Cabala," in *The Jewish Encyclopaedia*, vol. 3, p. 473.

42. Cf. Peter A. Redpath, "The Mystery of Israel: Antisemitism as an Enlightenment Metaphysical Principle," 2000 International Meeting American Maritain Association: "Faith, Scholarship, and Culture in the 21st Century," 19–22 October 2000, University of Notre Dame; later published under title "Anti-Semitism as an Enlightenment Metaphysical Principle," in *Contemporary Philosophy*, 23: 3, 4 (May/June & July/Aug, 2001).

Chapter Twenty-Four

1. Cf. Frank E. Manuel, Fritzie P. Manuel, *Utopian Thought in the Western World* (Cambridge, Mass.: The Belknap Press of Harvard University, 1979), p. 1.

2. Cf. P. M. Rattansi, "The Social Interpretation of Science in the Seventeenth Century," in *Science and Society 1600–1900*, ed. Peter Mathias (Cambridge, England: Cambridge University Press, 1972), p. 17.

3. Francis Bacon, *New Atlantis*, ed. Brian Vickers (Oxford, New York: Oxford University Press, 1996), p. 464; see also, Bacon, *New Atlantis*, The Internet Wiretap Edition (written in 1626), *From Ideal Commonwealths*, prepared by Kirk Crady (New York: P. F. Collier & Son, The Colonial Press, expired, 1901). This book is in the public domain, released August 1993, from scanner output provided by Internet Wiretap; see <kcrady@polaris.cv.nrao.edu>.

4. Ibid., p. 469.

5. Ibid., p. 480.

6. Ibid., pp. 480–488.

7. Francis Bacon, *La Nouvelle Atlantide*, trans. into French Michèle Le Doeuff, Margeret Llasera (Paris: Payod, 1983), p. 97.

8. Ibid., p. 110.

9. Cf. P. M. Rattansi, "The Social Interpretation of Science in the Seventeenth Century," in *Science and Society 1600–1900*, pp. 18–19, 27, 28.

10. William Eamon, "From the Secrets of Nature to Public Knowledge," in *Reappraisals of the Scientific Revolution*, ed. David C. Lindberg, Robert S. Westman, (Cambridge, England: Cambridge University Press 1990), p. 350; cf. Pierre Montloin, Jean-Pierre Bayard, *Les Rose-Croix* (Paris: Grasset, 1971), p. 75–76.

11. Bacon, *New Atlantis*, p. 462; Francis A. Yates, *The Occult Philosophy in the Elizabethan Age* (London, Boston: Henley, 1979), p. 174–175.

12. Charles Webster, *The Great Instauration: Science, Medicine, and Reform 1626–1660* (London: Duckworth, 1975), p. 50.

13. Ibid., p. 88.

14. Cf. Ernan McMullin, *Introduction*, in *The Social Dimension of Science*, ed. Ernan McMullin (Notre Dame, Ind.: 1992), p. 12 ff.

15. K. Theodore Hoppen, "The Early Royal Society," in *The British Journal for the History of Science*, 9:1 (March 1976), pp. 1–24n 31, 243–273n 32.

16. Ibid., n. 1, p. 2.
17. Ibid., p. 10.
18. Ibid., p. 12.
19. Richard H. Popkin, "The Religious Background of Seventeenth–Century Philosophy," in *Journal of the History of Philosophy*, 25:1 (January 1987), p. 47.
20. Hoppen, "The Early Royal Society," p. 13.
21. Popkin, "The Religious Background of Seventeenth–Century Philosophy, " p. 47.
22. Ibid., p. 14.
23. Ibid., p. 265.
24. Karl, R. H. Frick, *Die Erleuchteten:Gnostich-theosophische und alchemistisch-rosenkreuzerische Geheimgesellschaften bis zum Ende des 18. Jahrhunderts—ein Beitrag zur Geistesgeschichte der Neuzeit* [*The Illuminati: Gnostic-theosophic and Alchemic-Rosecrucian Secret Societies up to the End of the Eighteenth Century—A Contribution to the Spiritual History of Modern Times*] (Graz, Austria: Akademische Druck- u. Verlagsanstalt, 1973), p. 313.
25. Hoppen, "The Early Royal Society, " p. 13, passim.

Chapter Twenty-Five

1. Joseph Owens, *The Doctrine of Being in the Aristotelian "Metaphysics": A Study in the Greek Background of Mediaeval Thought* (Toronto: Pontifical Institute of Mediaeval Studies, 1978), p. 177.
2. Francis Bacon, *Novum organum* (New York: Collier, 1901), Polish trans. Jan Wikarjak (Warsaw: Państwowe Wydawnictwo Naukowe [State Scientific Publisher], 1955), 2. 2.
3. Ibid.
4. Mieczysław A. Krąpiec, Stanisław Kamiński, *Z teorii i metodologii metafizyki* [*From the Theory and Methodology of Metaphysics*] (Lublin: Redakcja Wydawnictw [Editorial Board of Publishers], Catholic University of Lublin, 1994), p. 220.
5. Bacon, *Novum organum*, 1.51.
6. Ibid., 2.2.
7. Ibid., 2.134.
8. Ibid., 2.2.
9. Ibid., 2. 8.
10. HerbertButterfield, *The Origins of Modern Science 1300–1800* (London: G. Bell and Sons, Ltd., 1949), p. 77.
11. Ibid., p. 79.
12. Bacon, *Novum organum*, 1.105.
13. Ibid., 1.14.
14. Michael Malherbe, "Bacon's Method of Science," in *The Cambridge Companion to Bacon*, ed. Markku Peltonen (Cambridge, England: Cambridge University Press, 1996), p. 78 ff.
15. Ibid., pp. 79–80.
16. Bacon, *Novum organum*, 1.105.
17. Bacon, *Novum organum*, Polish trans. Jan Wikarjak (Warsaw: Państwowe Wydawnictwo Naukowe, 1955), pp. 398–399nn 100, 101.
18. Ibid., p. 81.

19. Mieczysław A. Krąpiec, *Realizm ludzkiego poznania* (Poznań: Pallotinum, 1959), pp. 274–275.

20. William Eamon, "From the Secrets of Nature to Public Knowledge," in *Reappraisals of the Scientific Revolution*, ed. David C. Lindberg, Robert S. Westman (Cambridge, England: Cambridge University Press, 1990), p. 343 ff.

21. Charles Webster, *The Great Instauration: Science, Medicine and Reform 1626–1660* (London: Duckworth, 1975), p. 338.

22. P. M. Rattansi, *The Social Interpretation of Science in the Seventeenth Century*, in *Science and Society 1600–1900*, ed. P. Mathias (Cambridge, England: Cambridge University Press, 1972), pp. 11–12.

23. Ibid., p. 6.

Chapter Twenty-Six

1. Ugo Bianchi, "*Le Gnosticisme: Concept, Terminologie, Origines, Délimitation*" ["Gnosticism: Concept, Terminology, Origins, Delimitation"] in *Gnosis: Festschrift für Hans Jonas* [*Gnosis: Festschrift for Hans Jonas*], no editor given (Göttingen: Vandenhoeck & Ruprecht, 1978), p. 33.

2. Cf. Giovanni Filoramo, *A History of Gnosticism* (Cambridge, Mass.: Basil Blackwell, 1990), p. XIV.

3. Ibid., p. 1ff.

4. Plato, *Statesman*, 258 E–267 A; cf. Morton Smith, "The History of the Term '*Gnostikos*,'" in *The Rediscovery of Gnosticism*, ed. B. Layton (Leiden: E. J. Brill, 1981), vol. 2: *Sethian Gnosticism*, p. 798ff.

5. Ibid., p. 160.

6. Carsten Colpe," The Challenge of Gnostic Thought for Philosophy, Alchemy and Literature," in *The Rediscovery of Gnosticism*, vol. 1, p. 36; Giles Quispel, *Gnoza* [*Gnosis*], trans. into Polish B. Kita (Warsaw: Pax, 1988), p. 62.

7. *Genesis*, 1:31.

8. Filoramo, *A History of Gnosticism* , p. 159.

9. Quispel, *Gnoza*, p. 75.

10. Ibid., p. 93f f.

11. Filoramo, *A History of Gnosticism*, p. 148ff.

12. Wincenty Myszor, "Introduction" to Quispel, *Gnoza*, p. 32.

13. Ernst Benz, *Les sources mystiques de la philosophie romantique allemande* [*The Mystical Sources of the Romantic German Philosophy*] (Paris: J. Vrin, 1968), pp. 7–9.

14. Ibid., p. 13.

15. Ibid., p. 12.

16. Ibid., p. 14; cf. Zob B. Mojsisch, *Meister Eckhart. Analogie, Univozität, und Einheit* (Hamburg: 1983), p. 133; Karl Albert, *Meister Eckhart und die Philosophie des 14. Jahrhunderts* (Leipzig: 1993), p. 102.

17. Ibid., p. 17.

18. Boehme's basic works include *Morgenröte im Aufgang*, called *Aurora* (1612), *Die Drei Prinzipien göttlichen Wesens* (1619), *Vom dreifachen Leben des Menschen* (1619–20), *Vierzig Fragen von den Seelen* (1620), *Von der Menschwerdung Jesu Christi* (1620), *Sechs theosophische Punkte* (1620), *De signatura rerum* (1622), *Von der Gnadenwahl* (1623), *Mysterium magnum* (1623). Editions: August Faust, Will-Eric

252   SCIENCE IN CULTURE

Peuckert, *Sämtliche Schriften, 1941–1942, 1955–1961*, vols. 1–9; Werner Buddecke, *Die Urschriften, 1963–1966*, vols. 1–2.
19.   Robert F. Brown, *The Later Philosophy of Schelling: The Influence of Boehme on the Works of 1809–1815* (Lewisburg, Pa.: Bucknell University Press, 1977), passim.

Part Six

Chapter Twenty-Seven

1.   Emmet Kennedy, *Destutt de Tracy and the Origins of "Ideology"* (Philadelphia: The American Philosophical Society, 1978), p. X.
2.   Brian William Head, *Ideology and Social Science: Destutt de Tracy and French Liberalism* (Dordrecht Boston ,Lancaster: Martinus Nijhof Publishers, 1985), p. 27.
3.   Louis-Bertrand Geiger, *"Les idées divines dans l'oeuvre de S. Thomas,"* in *St. Thomas Aquinas 1274–1974, Commemorative Studies*, ed.-in-chief Armand A. Maurer (Toronto: Pontifical Institute of Mediaeval Studies, 1974), vol. 1, pp. 175–209.
4.   René Descartes, *Meditations on First Philosophy*, Bk. 3, 4–5.
5.   John Locke, "Introduction," *An Essay Concerning Human Understanding* (New York: Dover Publications, 1959), p. 8.
6.   Ibid., "Introduction."
7.   Ibid., p. 25.
8.   Head, *Ideology and Social Science: Destutt de Tracy and French Liberalism*, p. 10.
9.   Étienne Bonnot de Condillac, *Traité des sensations* (1754).
10.   Head, *Ideology and Social Science: Destutt de Tracy and French Liberalism*, p. 32.
11.   Ibid., p. 33.
12.   Brian William Head, *Politics and Philosophy in the Thought of Destutt de Tracy (1754–1836)* (New York; Graland Pub., 1987), pp. 76–77.
13.   Head, *Ideology and Social Science: Destutt de Tracy and French Liberalism*, p. 33.
14.   Ibid., p. 34.
15.   Kennedy, *Destutt de Tracy and the Origins of "Ideology,"* p. 45.
16.   Head, *Ideology and Social Science: Destutt de Tracy and French Liberalism*, p. 32ff; *Politics and Philosophy in the Thought of Destutt de Tracy (1754–1836)*, p. 86ff.
17.   Kennedy, *Destutt de Tracy and the Origins of "Ideology"*, p. 45.
18.   Head, *Ideology and Social Science: Destutt de Tracy and French Liberalism*, p. 35.
19.   Ibid., p. 35 ff.
20.   Head, *Politics and Philosophy in the Thought of Destutt de Tracy (1754–1836)*, p. 80.
21.   Ibid., p. 85.
22.   Head, *Ideology and Social Science: Destutt de Tracy and French Liberalism*, p. 37.

Chapter Twenty-Eight

1.   Francis Bacon, *Novum organum* (New York: Collier, 1901), Polish trans. Jan Wikarjak (Warsaw: Państwowe Wydawnictwo Naukowe [State Scientific Publisher], 1955), 1.34ff.
2.   Ibid.
3.   John Locke, "Introduction," *An Essay Concerning Human Understanding* (New York: Dover Publications, 1959), p. 4.
4.   Jorge Larrain, *The Concept of Ideology* (London: Hutchison, 1979), p. 22.

Chapter Twenty-Nine

1.   Brian William Head, *Ideology and Social Science: Destutt de Tracy and French Liberalism* (Dordrecht Boston, Lancaster: Martinus Nijhof Publishers, 1985), p. 54.
2.   Ibid.
3.   Ibid., p. 56.
4.   Ibid., p. 58.
5.   Ibid., p. 59.
6.   Ibid., p. 60.

Chapter Thirty

1.   Brian William Head, *Ideology and Social Science: Destutt de Tracy and French Liberalism* (Dordrecht, Boston, Lancaster: Martinus Nijhof Publishers, 1985), p. 10.
2.   Ibid., p. 11.

Chapter Thirty-One

1.   Jorge Larrain, *The Concept of Ideology* (London: Hutchison, 1979), p. 33.
2.   Ibid., p. 46.
3.   Ibid., p. 76ff.
4.   Ibid., pp. 100–130.
5.   Ibid., p. 83ff.
6.   Ibid., p. 88ff.
7.   Ibid., p. 91ff.

Chapter Thirty-Two

1.   Władysław Tatarkiewicz, *Historia Filozofii* [*History of Philosophy*] (Warsaw: Państwowe Wydawnictwo Naukowe [State Scientific Publisher], 1970), vol. 2, p. 70.
2.   Immanuel Kant, *Critique of the Pure Reason*, trans. J. M. D. Meiklejohn (New York: Collier Publishing Co., 1987) vol.1, pp. xi–xii.
3.   Ibid.. and Bk. 2, ch. 2, sect. 1.
4.   Ibid., Bk. 1, ch. 1, sect. 2.
5.   Ibid, Bk. 2, ch. 2, sect. 3.
6.   Mieczysław A. Krąpiec, *Filozofia—Co wyjaśnia?* [*Philosophy—What does It Explain?*] (Lublin: Lubelska Szkoła Filozofii Chrześcijańskiej [Lublin School of Christian Philosophy], 1999), p. 124.

7.   Kant, *Critique of the Pure Reason*, Bk. 1, ch. 2, sect. 2, ss 20; Ibid., ss 21.

8.   Ibid, p. 12 ff.

9.   Ibid., Bk. 2. Mieczysław A. Krąpiec, *Ja-człowiek*, *[I-Man]* (Lublin: Redakcja Wydawnictw [Editorial Board of Publishers] Katolicki Uniwersytet Lubelski [Catholic University of Lublin], 1991), p. 130 ff.

10.   Immanuel Kant, *Foundation of the Metaphysics of Morals* [*Uzasadnienie metafizyki moralności*, trans. Roman Ingarden (Warsaw: Państwowe Wydawnictwo Naukowe [State Scientific Publisher], 1984), sect. 2.

11.   Cf. Jacek W. Woroniecki, *Katolicka etyka wychowawcza* [*Catholic Educational Ethics*] (Lublin: Redakcja Wydawnictw [Editorial Board of Publishers] Katolicki Uniwersytet Lubelski [Catholic University of Lublin], 1988), vol. 1, p. 84.

12.   Martin Heidegger, *Pytanie o technikę* [*Question about Technology*], trans. into Polish K. Wolicki under the title *Budować, mieszkać, myśeć* [*To Build, To Live, To Think*] (Warsaw: Czytelnik, 1977), p. 240.

13.   Max Scheler, *Pisma z antropologii filozoficznej i teorii wiedzy* [*Writings in Philosophical Anthropology and the Theory of Knowledge*], Polish trans. Stanisław Czerniak and Adam Węgrzycki (Warsaw: Państwowe Wydawnictwo Naukowe[State Scientific Publisher], 1987), p. 339 ff.

Chapter Thirty-Three

1.   Auguste Comte, *Rozprawa o duchu filozofii pozytywnej: Rozprawa o całokszałcie pozytywizmu* [*Treatise on the Spirit of Positive Philosophy: Treatise on the Whole of Positivism*], originally published in French as *Discours sur l'ensemble du positivsime* [*Discourse on the Whole of Positivism*] (Paris, 1848), Polish trans. Barbara Skarga (Warsaw: Państwowe Wydawnictwo Naukowe [State Scientific Publisher],1973), pp. 45–49.

2.   Ibid., p. 246ff.

3.   Ibid., p. 64ff.

4.   Ibid., p. 168.

5.   Ibid., p. 273.

6.   Auguste Comte, *Metoda pozytywna w 16 wykładach* [*Positive Method in Sixteen Readings*], an abbreviated version of *Cours de philosophie positive*, vols. 1–6, (Paris, 1830–1842), Polish trans. Wanda Wojciechowska (Warsaw: Państwowe Wydawnictwo Naukowe, 1961), pp. 11–12.

7.   Ibid., p. 12.

8.   Ibid.,

9.   Comte, *Rozprawa o duchu filozofii pozytywnej: Rozprawa o całokszałcie pozytywizmu*, p. 54.

10.   Comte, *Metoda pozytywna w 16 wykładach*, p. 15.

11.   Ibid., p. 302.

12.   Ibid., p. 304.

13.   Ibid.

14.   Ibid., pp. 20–22.

15.   Ibid., p. 26.

16.   Ibid., p. 309.

17.   Ibid., p. 26.

18.   Ibid.

19. Ibid., p. 318.
20. Ibid., p. 339.
21. Ibid., p. 27.
22. Ibid., p. 28.
23. Ibid., p. 35.
24. Ibid., p. 39.
25. Ibid., p. 287.
26. Comte, *Rozprawa o duchu filozofii pozytywnej: Rozprawa o całokształcie pozytywizmu*, p. 111.
27. Comte, *Metoda pozytywna w 16 wykładach*, p. 292.
28. Ibid., p. 299.
29. Ibid., p. 335.
30. Ibid., p. 338.
31. Ibid., p. 317.
32. Ibid., p. 322.
33. Ibid., p. 331.
34. Ibid., p. 308.
35. Comte, *Rozprawa o duchu filozofii pozytywnej: Rozprawa o całokształcie pozytywizmu*, pp. 47, 268.
36. Comte, *Metoda pozytywna w 16 wykładach*, p. 345.
37. Christopher Dawson, *Postęp i religia* [*Progress and Religion*], trans. into Polish H. Bednarek (Warsaw: Państwowe Wydawnictwo Naukowe, 1958), pp. 189–215.
38. Eric Voegelin, *From Enlightenment to Revolution* (Durham, North Carolina: Duke University, 1975), pp. 76–79.
39. Ibid., p. 160ff.
40. Ibid., p. 166.
41. Ibid., p. 166ff.
42. Ibid., p. 136.

Chapter Thirty-Four

1. Hanna Buczyńska-Garewicz, *Koło Wiedeńskie* [*The Vienna Circle*] (Toruń: Comer, 1993), p. 9.
2. Władysław Tatarkiewicz, *Historia filozofii* [*History of Philosophy*] (Warsaw: Państwowe Wydawnictwo Naukowe[State Scientific Publisher],1970), vol. 3, pp. 99–100.
3. Byczyńska-Garewicz, *Koło Wiedeńskie*, p. 9.
4. Ibid., p. 11.
5. Ibid., pp. 47–49.
6. Ibid., p. 50–51.
7. Ibid., p. 59.
8. Ibid., pp. 54–55.

Chapter Thirty-Five

1. Bertrand Russell, *The Impact of Science on Society* (London: Allen & Unwin, 1952), p. 9.

2. Cf. Max Weber, *Etyka protestancka a duch kapitalizmu* [*The Protestant Ethic and the Spirit of Capitalism*], Polish trans. Jan Miziński (Lublin: Test, 1994).
3. Russell, *The Impact of Science on Society*, p. 119.
4. Zbigniew Brzeziński, *Between Two Ages* (New York: Viking Press, 1970), p. 9.
5. Russell, *The Impact of Science on Society*, p. 39.
6. Dante Alighieri, *De la monarchie* [*On Monarchy*] intro. and trans. into French B. Landry (Paris: 1933).
7. Russell, *The Impact of Science on Society*, p. 38.
8. Cf. Denise Laurence Cuddy, *The Road to Socialism and the New World Order* (Highland City, Florida: Florida Pro Family Forum, 1995), p. 2.
9. President George Herbert Walker Bush, Speech to the Nation, September 11, 1990. Cf. W. F. Jasper, *Global Tyranny . . . Step by Step* (Appleton, Wisc.: Western Island Publishers, 1992), p. IX.
10. Cited after Cuddy, *The Road to Socialism and the New World Order*, p. 23.
11. Zbigniew Brzeziński, *Wielka Szachownica* [*The Great Chessboard*], trans. into Polish by T. Wyżiński (Warsaw: Bertelsmann Media, 1999), p. 9.
12. Cf. Stanisław Wielgus, "Otwarcie Społecznego Tygodnia na jubileusz Metropolii Warszawskiej" ["Opening of the Social Week for the Jubilee of the Warsaw Metropolis"] (May 18, 2000), in *Człowiek w kulturze* [*Man and Culture*] (Lublin: Catholic University of Lublin, 2002), vol. 14.
13. Cf. Hans-Peter Martin, Harald Schumann, *Pułapka globalizacji. Atak na demokrację i dobrobyt* [*The GlobalizationTrap: Attack on Democracy and Prosperity*], English trans. Patrick Camiller, Polish trans. M. Zybura, from the original German work *Globaliserungsfalle* (New York: Zed Books, 1997).
14. Cf. Brzeziński, *Between Two Ages*, p. 74.
15. Cf. Cuddy, *The Road to Socialism and the New World Order*, p. 52.

Chapter Thirty-Six

1. Marcus Tullius Cicero, *Disputationes Tusculanae* Cambridge 1738, [*Tusculan Disputations*], in 5 books, ed. John Davis (London: J. & P. Knapton & Franc. Changuion, 1738), 2.13.
2. Cf. also Immanuel Kant, *Critique of the Pure Reason*, trans. J. M. D. Meiklejohn (New York: Collier, 1987). 2, ch. 2, sect. 1.
3. Cf. Edward Newton, *The Meaning of Beauty* (London: Penguin Books, 1962), p. 213.
4. Cf. Mieczysław A. Krąpiec, *Ja-człowiek*, [*I-Man*] (Lublin: Redakcja Wydawnictw [Editorial Board of Publishers], Catholic University of Lublin, 1991), pp. 157–177 and J. Chełstowski *Dylematy współczesnych badań biomedycznych—nadzieje i zagrożenia* [*Dilemnas of Modern Biomedical Research—Hopes and Dangers*] (Siedlce: Wydawnictwo Akademii Podlaskiej, 2000).

Chapter Thirty-Seven

1. Cf. Piotr Jaroszyński, *Metaphysics and Art*, trans. Hugh McDonald (New York: Peter Lang, 2002).

Notes 257

2. St. Thomas Aquinas, *Summa theologiae*, in *Sancti Thomae Aquinatis, doctoris angelici Opera Omnia iussu Leonis XIII. O.M. edita, cura et studio fratrum praedicatorum* (Rome: 1882–1996), 2–2, q. 4, a. 2 c.
3. Cf. Mieczysław A. Krąpiec, Stanisław Kamiński, *Z teorii i metodologii metafizyki* [*On the Theory and Methodology of Metaphysics*] (Lublin: Redacja Wydawnictw [Editorial Board of Publishers], Catholic University of Lublin, 1994), p. 15–17; Stanisław Kamiński, *Pojęcie nauki i klasyfikacja nauk* [*The Concept of Science and Classification of Sciences*] (Lublin: Catholic University of Lublin,1981), p. 11–18.

Chapter Thirty-Eight

1. Aristotle, *Metaphysics*, English trans. Hugh Tredennick (Cambridge Mass.: Harvard University Press, Loeb Classical Library, 1933); *Metafizyka*, Polish trans. Tadeusz Żeleźnik, ed. Mieczysław Albert Krąpiec, Andrzej Maryniarczyk (Lublin: Redakcja Wydawnictw Naukowych [Editorial Board of Publishers] Catholic University of Lublin, 1996), Bk. 1, ch. 1, 980a 22.
2. Mieczysław Albert Krąpiec, *O ludzką politykę* [*On a Human Politics*] (Warsaw: Guttenberg-Press, 1996), ch. 6, 2, 3.
3. Pope John Paul II, *Evangelium vitae* [*The Gospel of Life*], and *Solicitudo rei socialis* [*Concern for Social Reality*]. See also *Uniwersytety w nauczaniu Jana Pawła II* [*Universities in the Teaching of John Paul II*] (Warsaw: Szkoła Wyższa im. Bogdana Jańskiego [Bogdan Janski Higher School], 1999), vol. 1.

BIBLIOGRAPHY

Non-Polish Primary Sources:

Aquinas, St. Thomas. *Commentary on the Metaphysics of Aristotle.* Translated by J. P. Rowan, Chicago: Henry Regnery Co., 1961, volumes 1, 2.

―――. *Expositio super librum Boethii de Trinitate,* in *Opusculae theologiae.* Edited by Marietti (Turin, Rome: 1954), vol. 2

―――. *In I Sententiarum, Scriptum super libros Sententiarum.* Edited by Pierre Mandonnet. Paris: 1929, volume 1.

―――. *In III Sententiarum, Scriptum super libros Sententiarum.* Edited by M. F. Moos. Paris: 1933, volume 3.

―――. *Quaestiones disputatae de veritate,* in *Sancti Thomae Aquinatis, doctoris angelici Opera Omnia iussu Leonis XIII. O.M. edita, cura et studio fratrum praedicatorum.* Rome: 1882–1996, volume 22.

―――. *Sancti Thomae Aquinatis, doctoris angelici Opera Omnia iussu Leonis XIII. O.M. edita, cura et studio fratrum praedicatorum.* Rome: 1882–1996.

―――. *Summa theologiae,* in *Sancti Thomae Aquinatis, doctoris angelici Opera Omnia iussu Leonis XIII. O.M. edita, cura et studio fratrum praedicatorum.* Rome: 1882–1996, volumes 4–11.

Aristotle. *Aristotelis opera.* Academia regia Borussica: Berolini 1837–1870, t. 1–5.

―――. *The Categories, On Interpretation.* Translated by Harold P. Cooke. Cambridge, Mass.: Harvard University Press, Loeb Classical Library, 1938.

―――. *Metaphysica.* Edited by Werner Jaeger. Oxford: The Clarendon Press, 1963.

―――. *Metaphysics.* Translated by Hugh Tredennick. Cambridge Mass.: Harvard University Press, 1933, Loeb Classical Library, in two volumes.

―――. *Nicomachean Ethics.* Translated by Harris Rackham. Cambridge, Mass.: Harvard University Press, Loeb Classical Library, 1934.

―――. *On the Heavens.* Translated by W. K. C. Guthrie. Cambridge, Mass.: Harvard University Press, Loeb Classical Library, 1960.

―――. *Physics.* Translated by Philip H. Wicksteed and Francis M. Cornford. Cambridge, Mass.: Harvard University Press, Loeb Classical Library, 1960–1963.

―――. *Politics.* Translated by Harris Rackham. Cambridge, Mass.: Harvard University Press, Loeb Classical Library, 1932.

Aristotle. *Posterior Analytics*. Translated by Hugh Tredennick, in Hugh Tredennick and E. S. Forster (trans.), *Posterior Analytics. Topica*. Cambridge Mass.: Harvard University Press, Loeb Classical Library,1934.

————. *Topica*. Translated by E. S. Forster, in Hugh Tredennick and E. S. Forster (trans.), *Posterior Analytics. Topica*. Cambridge Mass.: Harvard University Press, Loeb Classical Library,1934.

Augustine, St. Aurelius. *De praedestinatione sanctorum* [*On the Predestination of the Blessed*], ii, in *Patrologia latina*. Edited by Migne. Paris: 1978–1990, vol. 44.

————. *Epistola*, in *Patrologia latina*, ed. Migne. Paris: 1978–1990, vol. 44.

————. *Sermo* [*Sermon*], 37, 5, 6, " *New Advent*, URL= http://www.newadvent.org/fathers/160337.htm

Bacon, Francis. *Advancement of Learning, New Atlantis*. Selected and edited by Richard Foster Jones. New York: Odyssey Press, 1937.

————. *New Atlantis*. Edited by Brian Vickers (ed.). Oxford, New York: Oxford University Press, 1996.

————. *The Great Instauration*, in *The English Philosophers from Bacon to Mill*. Edited by E. A. Burtt. New York: Random House, 1939.

————. *Novum organum*. New York: Collier Publishing Co., 1901.

————. *La Nouvelle Atlantide* [*New Atlantis*]. Translated into French by Michèle Le Doeuff and Margaret Llasera. Paris: Payod, 1983.

Cicero, Marcus Tullius. *De re publica, De legibus*. Translated by Clinton Walker Keyes. Cambridge Mass.: Harvard University Press, Loeb Classical Library, 1988.

————. *Disputationes Tusculanae* Cambridge 1738 [*Tusculan Disputations*], in 5 books. Edited byJohn Davis. London: J. & P. Knapton & Franc. Changuion, 1738.

Condillac, Etienne Bonnot de. *Oeuvres de Condillac, rev. cor. par l'auteur, imprimées sur ses manuscrits autographes, et augm. de La langue des calculs, ouvrage posthume*. Paris: C. Houel, 1798.

Clement of Alexandria, *Protrepticus und Paedagogus*. Edited by Otto Stälin. Leipzig: J. C. Hinrichs, 1905–1909.

Damian, St. Peter. *De divina omnipotentia et altri opuscoli* [*Concerning Divine Omnipotence and Other Works*]. Edited by Paolo Brezzi. Translated by Bruno Nardi. Firenze, Vallecchi: 1943.

Dante (Alighieri). *De la monarchie*. Introduction and translation into French by B. Landry. Paris: Felix Alcan, 1933.

Descartes, René. *Discourse on Method and Meditations*. Translated by Laurence J. Lafleur. New York: Liberal Arts Press, 1960.

Diels, Hermann. *Die fragmente der Vorsokratiker, griechisch und deutsch* [*The Fragments of the Presocratics: Greek and German*]. Berlin: Weidmannsche Buchhandlung, 1906–1910.

Diogenes Laertius. *Lives and Views of Eminent Philosophers*. Translated by Robert Drew Hicks. Cambridge Mass.: Harvard University Press, Loeb Classical Library, 1958–1959.

Heinze, Richard. *Xenokrates. Darstellung der Lehre und Sammlung der Fragmente*. Hildesheim: Georg Olms Verlagsbuchhandlung, 1965.

Herodotus of Harlicarnassus. *Herodotus: The Histories, Newly Translated*. No translator given. Harmondsworth, Midx., England: Penguin Books, 1954.

Hippocrates. *Works*. English translation by W. H. S. Jones. Cambridge: Harvard University Press, Loeb Classical Library, 1959–1962.

Kant, Immanuel. *Critique of Pure Reason*. Translated by J. M. D. Meiklejohn. New York: Collier Publishing Co., 1787.

Locke, John. *An Essay Concerning Human Understanding*. New York: Dover Publications, 1959.

More, Thomas. *Utopia*. Translated by Raphe Robynson. Cambridge, England: Cambridge University Press, 1956.

Pico della Mirandola, Giovanni. *Oration on the Dignity of Man*. Translated by A. Robert Caponigri (Chicago: Regnery Publishing, 1956); URL=http://en.wikipedia.org/wiki/Giovanni_Pico_della_Mirandola.

Plato. *The Laws*. Translated by R. G. Bury. Cambridge Mass.: Harvard University Press, 1961.

———. *Lysis, Symposium. Gorgias*. Translated by Walter R. M. Lamb, Cambridge Mass.: Harvard University Press, Loeb Classical Library, 1961.

———. *Euthyphro, Apology, Crito, Phaedo, Phaedrus*. Translated by Harold North Fowler. Cambridge Mass.: Harvard University Press, Loeb Classical Library, 1914.

———. *Platonis opera*. Edited by John Burnet. Oxford: The Clarendon Press, 1900–1907; repr. 1950, 5 volumes.

———. *Timaeus, Critias, Cleitophon, Menexenus, Letters*. Translated by R. G. Bury. Cambridge Mass.: Harvard University Press, Loeb Classical Library, 1961.

Plotinus. *Enneads with an English Translation by A.H. Armstrong*. Translated by Arthur Hilary Armstrong. Cambridge Mass.: Harvard University Press, Loeb Classical Library, 1966.

Quintilian, Marcus Fabius. *The Institutes of Oratory* [*Institutio oratoria*]. Translated by Harold Edgeworth Butler. Cambridge, Mass.: Harvard University Press, Loeb Classical Library, 1959–1963.

Seneca, Lucius Annaeus. *Ad Lucilium epistulae morales* [*Moral Epistles to Lucilius*]. *Seneca* ; *with an English Translation by Richard M. Gummere*. Translated by Richard M. Gummere. Cambridge, Mass.: Harvard University Press, Loeb Classical Library, 1996.

Polish Primary Sources

Arystoteles (Aristotle). *Analityki pierwsze i wtóre* [*Prior and Posterior Analytics*]. Translated by Kazimierz Leśniak. Warsaw: Państwowe Wydawnictwo Naukowe [State Scientific Publisher], 1973.

Kategorie i Hermeneutyka z dodaniem Isagogi Porfiriusza [*Categories and Hermeneutics with the addition of Porphyry's Isagoge*]. Translated into Polish by Kazimeirz Leśniak. Warsaw: Państwowe Wydawnictwo Naukowe [State Scientific Publisher], 1973.

―――. *Metafizyka*. Translated into Polish by Tadeusz Żeleźnik. Edited by Mieczysław Albert Krąpiec, Andrzej Maryniarczyk. Lublin: Redakcja Wydawnictw Naukowych [Editorial Board of Publishers], Catholic University of Lublin, 1996.

―――. *O niebie* [*On the Heavens*]. Translated into Polish by Pawel Siwek. Warsaw: Państwowe Wydawnictwo Naukowe [State Scientific Publisher], 1980.

―――. *Polityka* [*Politics*]. Translated into Polish by Lodovici Piotrowicz. Warsaw: Państwowe Wydawnictwo Naukowe [State Scientific Publisher], 1964.

―――. *Zachęta do filozofii* [*Exhortation to Philosophy*, also known as *Protrepticus*]. Translated into Polish by Kazimierz Leśniak. Warsaw: Państwowe Wydawnictwo Naukowe [State Scientific Publisher], 1988.

Bacon, Francis. *Novum Organum*. Translated into Polish by Jan Wikarjak. Warsaw: Państwowe Wydawnictwo Naukowe [State Scientific Publisher], 1955.

Cochrane, Charles Norris. *Chrześcijaństwo i kultura antyczna* [*Christianity and Classical Culture*]. Translated into Polish by Gabriela Pianko. Warsaw: Pax, 1960.

Comte, Auguste. *Metoda pozytywna w 16 wykładach* [*Positive Method in 16 Lectures*]. Translated ito Polish by Wanda Wojciechowska. Warsaw: Państwowe Wydawnictwo Naukowe [State Scientific Publisher], 1961; an abbreviated version of *Cours de philosophie positive* 1–6. Paris: 1830–1842.

Comte, Auguste. *Rozprawa o duchu filozofii pozytywnej. Rozprawa o całokształcie pozytywizmu*, [*Treatise on the Spirit of Positive Philosophy:Treatise on the Whole of Positivism*]. Translated into Polish by Barbara Skarga, Warsaw: Państwowe Wydawnictwo Naukowe [State Scientific Publisher], 1973; originally *Discours sur l'ensemble du positivisme* [*Discourse on the Whole of Positivism*], Paris 1848.

Condillac, Étienne Bonnot de. *Traktat o wrażeniach* [*Treatise on Sensations*]. Translated into Polish by Wanda Wojciechowska. Warsaw: Państwowe Wydawnictwo Naukowe [State Scientific Publisher], 1958.

Cyceron (Cicero). *O państwie* [*On the Republic*]. Translated into Polish by Wictor Kornatowski. Warsaw: Państwowe Wydawnictwo Naukowe [State Scientific Publisher], 1960.

Descartes, René. *Medytacje o pierwszej filozofii wraz z zarzutami uczonych mężów i odpowiedzią autora oraz rozmowa z Burmanem* [*Meditations on First Philosophy together with Objections of Erudite Men and the Author's Response, and a Conversation with Burman*]. Translated into Polish by Kazimierz and Maria Ajdukiewicz, Stefan Swieżawski. Warsaw: Państwowe Wydawnictwo Naukowe [State Scientific Publisher], 1958, volumes 1–2.

Kant, Immanuel. *Krytyka czystego rozumu* [*Critique of Pure Reason*]. Translated into Polish by Roman Ingarden. Warsaw: Państwowe Wydawnictwo Naukowe [State Scientific Publisher], 1957, volumes 1–2.

———. *Uzasadnienie metafizyki moralności* [*Foundation of the Metaphysics of Morals*]. Translated by into Polish by Roman Ingarden. Warsaw: Państwowe Wydawnictwo Naukowe [State Scientific Publisher], 1984.

Kwintylian [Quintilian, Marcus Fabius]. *Kształcenie mówcy* [*Formation of the Speaker*]. Translated into Polish by Mieczyslaw Brożek. Wrocław: Ossolineum 1951.

More, St. Thomas. *Utopia.* Translated into Polish by Kazimierz Abgarowicz, Warsaw: Pax, 1954.

Platon [Plato]. *Fajdros* [*Phaedrus*]. Translated into Polish by Leopold Regner. Warsaw: Państwowe Wydawnictwo Naukowe [State Scientific Publisher], 1993.

———. *Fedon* [*Phaedo*]. Translated into Polish by Władysław Witwicki. Warsaw: Państwowe Wydawnictwo Naukowe [State Scientific Publisher], 1982.

———. *Listy* [*Letters*]. Translated into Polish by Maria Maykowska. Warsaw: Państwowe Wydawnictwo Naukowe [State Scientific Publisher], 1987.

———. *Prawa* [*Laws, The*]. Translated into Polish by Władysław Witwicki. Warsaw: Państwowe Wydawnictwo Naukowe [State Scientific Publisher], 1958.

Platon [Plato]. *Timajos* [*Timaeus*]. Translated into Polish by Pawel Siwek. Warsaw: Państwowe Wydawnictwo Naukowe [State Scientific Publisher], 1986.

―――. *Uczta* [*Symposium*]. Translated into Polish by Władysław Witwicki. Warsaw: Państwowe Wydawnictwo Naukowe [State Scientific Publisher], 1975.

Plotyn [Plotinus]. *Enneady* [*Enneads*]. Translated into Polish by Adam Krokiewicz. Warsaw: Państwowe Wydawnictwo Naukowe [State Scientific Publisher], 1959.

Non-Polish Secondary Works

Albert, Karl. *Meister Eckhart und die Philosophie des 14. Jahrhunderts*. Leipzig: 1993.

Alexandrini, Philonis. *De mutatione nominum* [*On Change of Names*], in *Opera quae supersunt*. Berlin 1896–1926, vol. 3.

Aubenque, Pierre. *Le problème de l'etre chez Aristote* [*The Problem of Being According to Aristotle*]. Paris, Presses Universitaires de France, 1977.

Baudoux, Bernardus. "'*Philosophia: Ancilla Theologiae*" ["Philosophy: Handmaiden to Theology"], in *Antonianum*, Rome: Annus 12, 1937.

Baynes, Norman H., Moss, L. B.. *Byzantium: An Introduction to East Roman Civilization*. Oxford: The Clarendon Press, 1949.

Becco, Anne. "Leibniz et François-Mercure von Helmont: Bagatelle pour des monades," in *Magia naturalis und die Entstehung der modernen Naturwissenschaften. Symposion der Leibniz-Gesellschaft* [*Magia naturalis and the Emergence of the Natural Sciences. Symposium of the Leibniz Society*], Hanover, 14–15. November 1975. Edited by Albert Heinekamp and Dieter Mettler. Wiesbaden: Franz Steiner Verlag, 1978.

Benz, Ernst. *Les sources mystiques de la philosophie romantique allemande* [*The Mystical Sources of Romantic German Philosophy*]. Paris: J. Vrin, 1968.

Bernal, Martin. *Black Athena. The Afroasiatic Roots of Classical Civilization*. London: Free Association Books, 1987.

Bevan, Edwin. *Stoics and Sceptics*. New York: Amo Press, 1979.

Bréhier, Émile. *Histoire de la philosophie* [*History of Philosophy*]. Paris: Libraire Félix Alcan, 1931.

―――. *Chrysippe et l'ancien stoïcisme* [*Chrysippus and Ancient Stoicism*]. Paris: Presses Universitaires de France, 1951.

Bianchi, Ugo. "*Le Gnosticisme: Concept, Terminologie, Origines, Délimitation*," in *Gnosis. Festschrift für Hans Jonas* [*Gnosis: Festschrift for Hans Jonas*]. No editor given. Göttingen: Vandenhoeck & Ruprecht, 1978.

Brown, Robert F.. *The Later Philosophy of Schelling. The Influence of Boehme on the Works of 1809–1815*. Lewisburg, Pa.: Bucknell University Press, 1977.

Brzeziński, Zbigniew. *Between Two Ages*. New York: Viking Press, 1970.

Burnet, Johm. *Early Greek Philosophy*, New York: Meridian Books, 1930.

Burtt, E. A. (ed.). *The English Philosophers from Bacon to Mill*. New York: Random House, 1939.

Butterfield, Herbert. *The Origins of Modern Science 1300–1800*. London: G. Bell and Sons, Ltd., 1949

Cassirer, Ernst. *The Logic of the Humanities*. New Haven and London: Yale University Press, 1966.

Chenu, Marie-Dominique. *La theologie comme science au XIIIe siecle* [*Theology as Science in the 13th Century*]. Paris: J Vrin, 1957.

Chroust, Anton-Herman. "The Origin of Metaphysics," in *The Review of Metaphysics*, 14 (1960), pp. 601–617.

Cochrane, Charles Norris. *Christianity and Classical Culture: A Study of Thought and Action from Augustus to Augustine*. Oxford : The Clarendon Press, 1940.

Colish, Marcia L.. *The Stoic Tradition from Antiquity to the Early Middle Ages*. Leiden: E. J. Brill, 1985.

Colpe, Carsten. "The Challenge of Gnostic Thought for Philosophy, Alchemy, and Literature," in *The Rediscovery of Gnosticism*. Edited by B. Layton. Leiden: E. J. Brill, 1978, volume 1.

Copenhaver, Brian P.. "Hermes Trismegistus and Early Modern Science: The 'Yates Thesis,'" in *Reappraisals of the Scientific Revolution*, edited by David C. Lindberg, Robert S. Westman. Cambridge, England: Cambridge University Press, 1990.

Crombie, Alistair Cameron. *Robert Grosseteste and the Origins of Experimental Science 1100-1700*. Oxford: The Clarendon Press, 1953.

———. *Science in the Later Middle Ages and Early Modern Times: 13th to 17th Centuries*. Harmondsworth, Penguin, 1969.

Cuddy, Dennis Laurence. *The Road to Socialism and the New World Order*. Highland City, Florida: Florida Pro Family Forum, 1995.

Deely, John N.. *Four Ages of Understanding. The First Postmodern Survey of Philosophy from Ancient Times to the Turn of the Twenty-First Century*. Toronto, Buffalo, London: University of Toronto Press, 2001.

Des Places, E. (ed. and trans.). *Jamblique: Les mystères d'Egypte*. Paris : Les Belles Lettres, 1966.

Duhem, Pierre. *Le système du monde* [*The World System*]. Paris: A. Hermann, 1913–1956, volumes 1–7.

Eamon, William. "From the Secrets of Nature to Public Knowledge," in *Reappraisals of the Scientific Revolution*. Edited by David C. Lindberg, Robert S. Westman, Cambridge, England: Cambridge University Press, 1990.

Easton, Stewart C.. *Roger Bacon and his Search for a Universal Science*. Westport, Conn.: Greenwood Press, 1970.

Festugière, André-Jean. *La Révélation d'Hermes Trismegiste*.[*The Revelation of Hermes Trismegistus*]. Paris: Lecoffre, 1950–1954, volumes 1–4.

Filoramo, Giovanni. *A History of Gnosticism*. Cambridge, Mass.: Basil Blackwell, 1990.

Fowden, Garth. *The Egyptian Hermes: A Historical Approach to the Late Pagan Mind*. Princeton, New Jersey: Princeton University Press, 1986.

Frick, Karl R. H.. *Die Erleuchten Gnostisch-theosophische und alchemistisch-rosenkreuzerische Geheimgesellschaften bis zum Ende des 18. Jahrhunderts—ein Beitrag zur Geistesgeschichte der Neuzeit* [*The Illuminati: Gnostic-theosophic and Alchemic-Rosecrucian Secret Societies up to the End of the Eighteenth Century—A Contribution to the Spiritual History of Modern Times*], Graz, Austria: Akademische Druck- u. Verlagsanstalt, 1973.

Geiger, Louis-Bertrand. "*Les idées divines dans l'oeuvre de S.Thomas*" ["*Divine Ideas in the Work of St. Thomas*"], in *Saint Thomas Aquinas, 1274–1974, Commemorative Studies*. Editor-in-chief Armand A. Maurer. Toronto: Pontifical Institute of Mediaeval Studies, 1974, volume 1, pp. 175–209.

Gilson, Étienne. *Études de philosophie médiévale* [*Studies of Medieval Philosophy*]. Strasbourg: Commission des publications de la faculté des lettres Palais de l'Université de Strasbourg, 1921.

———. *History of Christian Philosophy in the Middle Ages*. New York: Random House, 1955.

Hackett, Jeremiah. "Roger Bacon on *Scientia Experimentalis*," in *Roger Bacon and the Sciences*. Edited by Jeremiah Hackett. Leiden, New York, Köln: E. J. Brill, 1997.

Hadot, Pierre. "Philosophie, discours philosophique, et divisions de la philosophie chez les stoiciens" ["Philosophy, Philosophic Discourse, and Stoic Divisions of Philosophy"], in *Revue international de la philosophie*, 45:178 (March 1991).

Head, Brian William. *Ideology and Social Science. Destutt de Tracy and French Liberalism*. Dordrecht, Boston, Lancaster: Martinus Nijhof Publishers, 1985.

Head, Brian William. *Politics and Philosophy in the Thought of Destutt de Tracy (1754-1836)*. New York: Garland Pubishing, 1987.

Heidegger, Martin. *The Question Concerning Technology, and Other Essays*. Translated by William Lovitt, New York: Harper & Row, 1977.

Heinekamp, Albert and Dieter Mettler (eds). *Magia naturalis und die Entstehung der modernen Naturwissenschaften. Symposion der Leibniz-Gesellschaft [Magia naturalis and the Emergence of the Natural Sciences. Symposium of the Leibniz Society]*, Hanover, 14–15. November 1975. Wiesbaden: Franz Steiner Verlag, 1978.

Hoppen, K. Theodore. "The Early Royal Society," in *The British Journal for the History of Science*, 9:31 (March 1976), pp. 1–24; 9:32 (June 1976), pp. 243–273.

Hülser, Karlheinz. *Die Fragmente zur Dialektik der Stoiker*. Stuttgart–Bad Cannstatt: Frommann-Holzboog, 1987.

Jaeger, Werner. *The Theology of the Early Greek Philosophers*. London, Oxford, New York: Oxford University Press, 1968; The Gifford Lectures, 1936.

Jaroszyński, Piotr. *Metaphysics and Art*. Translated by Hugh McDonald, New York: Peter Lang Publishing, 1999.

Jasper, W. F. *Global Tyranny . . . Step by Step*. Appleton, Wisc.: Western Islands Publishers, 1992.

Jewish Encyclopedia, The. New York and London, 1902, volume 3.

John Paul II, Pope. *On Human Work: Laborem Exercens*. U. S. Catholic Conference, Washington, D. C.: Hunter Publishing, 1981.

———. *Laborem exercens*. Wrocław: Wrocławskie Wydawnictwo Oświatowe, 1992.

Kennedy, Emmet. *Destutt de Tracy and the Origins of "Ideology."* Philadelphia: The American Philosophical Society, 1978.

Klemm, Friedrich. *A History of Western Technology*, Cambridge, Mass.: Harvard University Press, 1964.

Krąpiec, Mieczysław A.. *Metaphysics: An Outline of the History of Being*. Translated by Theresa Sandok. New York: Peter Lang, 1991.

Largeault, Jean. *Enquête sur le nominalisme [Inquiry on Nominalism]*. Paris: Béatrice-Nauwelaerts, Louvain: Éditions Nauwelaerts, 1971.

Larrain, Jorge. *The Concept of Ideology*. London: Hutchison London, 1979.

Layton, B. (ed.). *The Rediscovery of Gnosticism*. Leiden: E. J. Brill, 1981.

Lewy, Hans. *Chaldean Oracles and Theurgy: Mysticism, Magic, and Platonism in the Later Roman Empire*. Paris: Études Augustiniennes, 1978.

Lindberg, David C.. *The Beginnings of Western Science*. Chicago and London: University of Chicago Press, 1992.

———. Robert S. Westman (eds.). *Reappraisals of the Scientific Revolution*. Cambridge, England: Cambridge University Press, 1990.

Malherbe, Michael. "Bacon's Method of Science," in *The Cambridge Companion to Bacon*. Edited by Markku Peltonen. Cambridge: Cambridge University Press, 1996.

Manuel, Frank E. and Manuel, Fritzie P. *Utopian Thought in the Western World*. Cambridge, Mass.: The Belknap Press of Harvard University, 1979.

Maritain, Jacques. *Three Reformers: Luther—Descartes— Rousseau*. London: Sheed & Ward, 1950.

Mates, Benson. *Stoic Logic*. Berkeley, Los Angeles, London: Library Reprint Series, 1973.

Maurer, Armand A. (ed.-in-chief). *St. Thomas Aquinas, 1274–1974, Commemorative Studies*. Toronto: Pontifical Institute of Mediaeval Studies, 1974, volume 1, pp. 175–209.

McKeon, Richard (ed.). *Introduction to Aristotle*. New York: Modern Library, 1947.

McMullin, Ernan. "The Goals of Natural Science," in *Proceedings and Addresses of The American Philosophical Association*, 58:1 (1984).

———. *The Inference that Makes Science*. Marquette, Milwaukee: Marquette University Press, 1992.

——— (ed.). *The Social Dimension of Science*. Notre Dame, Ind.: University of Notre Dame Press, 1992.

Merlan, Philippe. *From Platonism to Neoplatonism*. The Hague 1960.

———. "On the Terms 'Metaphysics' and 'Being-qua-Being,'" in *Monist*, 52 (1968).

Mojsisch, Zob B.. *Meister Eckhart. Analogie, Univozität, und Einheit*. Hamburg: 1983.

Montloin, Pierre and Bayard, Jean-Pierre. *Les Rose-Croix*. Paris: Grasset, 1971.

Newton, Edward. *The Meaning of Beauty*. London: Penguin Books, 1962.

Open University, The: An Arts Foundation Course A 102 Units 31–32 Conclusion, Walton Hall: Milton Keynes, 1987.

Owens, Joseph. *The Doctrine of Being in the Aristotelian "Metaphysics."* Toronto: Pontifical Institute of Mediaeval Studies, 1978.

Pohlenz, Max. *Die Stoa. Geschichte einer geistigen Bewegung.* Göttingen: Vandenhoeck und Ruprecht, 1964.

Popkin, Richard H.. "The Religious Background of Seventeenth-Century Philosophy," in *Journal of the History of Philosophy*, 25:1 (1987).

Rattansi, P.M.. "The Social Interpretation of Science in the Seventeenth Century," in *Science and Society 1600-1900*, edited by Peter Mathias, Cambridge: University Press, 1972.

Redpath, Peter A.. "The Mystery of Israel: Antisemitism as an Enlightenment Metaphysical Principle," *2000 International Meeting American Maritain Association: Faith, Scholarship, and Culture in the 21st Century*, 19–22 October 2000, University of Notre Dame.

———. "Anti-Semitism as an Enlightenment Metaphysical Principle," in *Contemporary Philosophy*, 23: 3, 4 (May/June & July/Aug, 2001).

———. *Wisdom's Odyssey from Philosophy to Transcendental Sophistry.* Amsterdam and Atlanta: Value Inquiry Book Series [VIBS], Editions Rodopi, B.V., 1997.

Régis, Louis.-Marie. *L'Opinion selon Aristote [Opinion According to Aristotle].* Paris: J. Vrin; Ottawa: Institut d'études médiévales, 1935.

Rodier, Georges. *Études de philosophie grecque [Studies of Greek Philosophy].* Paris: J. Vrin, 1969.

Ross, William David. *Aristotle's Metaphysics.* Oxford: The Clarendon Press, 1981.

Rossi, Pado. *Francis Bacon: From Magic to Science.* Chicago: The University of Chicago Press, 1978.

———. "Bacon's Idea of Science," in *The Cambridge Companion to Bacon.* Edited by Markku Peltonen. Cambridge: Cambridge University Press, 1996.

Russell, Bertrand, *The Impact of Science on Society.* London: Allen & Unwin, 1959.

Sextus Empricus, *Adversus mathematicos*, 7.16, in *Sextus Empiricus.* Translated by R. G. Bury. Cambridge, Mass.: Harvard University Press, The Loeb Classical Library, 1939–1949, volumes 1–4.

Shaw, Gregory. *Theurgy and the Soul: The Neoplatonism of Iamblichus.* Pennsylvania: The Pennsylvania State University Press, 1995.

Smith, Morton. "The History of the Term '*Gnostikos*,'" in *The Rediscovery of Gnosticism.* Edited by B. Layton. Leiden: E. J. Brill, 1981, volume 2: *Sethian Gnosticism.*

Spannent, Michel. *Permanence du stoicisme: De Zenon a Malraux* [*The Permanence of Stoicism: From Zeno to Malraux*). Gembloux: Éditions J. Duculot, S. A., 1973.

"Tatian," in *The Catholic Encyclopedia*. New York: 1912, volume14.

Thorndike, L. *A History of Magic and Experimental Science*. New York: Columbia University Press, 1923–1958, volumes 1–8.

Timothy, Hamilton Baird. *The Early Christian Apologists and Greek Philosophy*. Assen: Van Gorcum, 1973.

Verbeke, Gerard. *"Philosophie et semiologie chez les stoïciens"* ["Philosophy and Semiology in the Stoics"], in *Études philosophiques présentées au Dr. Ibrahim Madkour* [*Philosophical Studies Presented to Dr. Ibrahim Madkour*], no editor given. Cairo, General Egyptian Book Organization, 1974.

————. *"[P]hilosophie du signe chez les stoïciens, La"* [" The Philosophy of the Sign in the Stoics"], in *Les stoiciens et leur logique* [*The Stoics and their Logic*]. Edited by J. Brunschwig. Paris: J. Vrin, 1978.

Voegelin, Eric. *From Enlightenment to Revolution*. Durham, North Carolina: Duke University, 1975.

Walker, Daniel Pickering. *The Ancient Theology: Studies in Christian Platonism from the Fifteenth to the Eighteenth Century*. London: Duckworth, 1972.

Wallis, Richard T.. *Neoplatonism*. London: Duckworth 1972.

Webster, Charles. *The Great Instauration: Science, Medicine, and Reform 1626–1660*. London: Duckworth, 1975.

Werner, Charles. *La philosophie grecque* [*Greek Philosophy*]. Paris: Payot, 1972.

Whewell, William. *History of the Inductive Sciences from the Earliest Times to the Present Time*. New York: Dover, 1959.

Whittaker, Thomas. *The Neo-Platonists: A Study in the History of Hellenism*. Hildesheim, Zürich, New York: Georg Olms, 1987.

Yates, Francis A. *Giordano Bruno et la tradition hermetique* [*Giordano Bruno and the Hermetic Tradition*]. Paris: Dervy-Livres, 1988.

————. *The Occult Philosophy in the Elizabethan Age*. London, Boston: Henley: Routledge & Kegan Paul, 1979.

Polish Secondary Works

Achmanow, Aleksander. *Logika Arystotelesa* [*Aristotle's Logic*]. Translated into Polish by A. Zabłudowski and Barbara Stanosz, Warsaw: Państwowe Wydawnictwo Naukowe [State Scientific Publisher], 1965.

Brzeziński, Zbigniew. *Wielka szachownica* [*The Great Chessboard*]. Translated into Polish by Tomasz Wyżyński. Warsaw: Bertelsmann Media, 1999.

Buczyńska-Garewicz, Hanna. *Koło Wiedeńskie* [*The Vienna Circle*]. Toruń: Comer, 1993.

Bugaj, Roman. *Hermetyzm* [*Hermeticism*]. Wrocław, Warsaw, Kraków: Ossolineum, 1991.

Chełstowski, Jerzy. *Dylematy współczesnych badań biomedycznych—nadzieje i zagrożenia* [*Dilemmas of Contemporary Biomedical Research—Hopes and Dangers*]. Siedlce: Wydawnictwo Akademii Podlaskiej, 2000.

Crombie, Alistair Cameron. *Nauka średniowieczna i początki nauki nowożytnej* [*Medieval Science and the Beginnings of Modern Science*]. Translated into Polish by S. Łypacewicz. Warsaw: Pax, 1960.

————. *Style myśli naukowej w początkach nowożytnej Europy* [*Styles of Scientific Thought in the Beginnings of Modern Europe*]. Translated into Polish by Piotr Salwa. Warsaw: Państwowe Wydawnictwo Naukowe [State Scientific Publisher], 1994.

Dawson, Christopher. *Postęp i religia* [*Progress and Religion*]. Translated into Polish by H. Bednarek. Warsaw: Państwowe Wydawnictwo Naukowe, 1958.

Domański, Julius. *Metamorfozy pojęcia filozofii* [*Metamorphoses of the Concept of Philosophy*]. Translated into Polish by Z. Mroczkowska, Mirosław Bujko, Warsaw: Instytut Filozofii i Socjologii, 1996.

Heidegger, Martin. *Pytanie o technikę* [*Question Concerning Technology*]. Translated into Polish by K. Wolicki. Warsaw: Czytelnik, 1977.

Jaeger, Werner. *Paideia*. Translated into Polish byMarian Plezia, Warsaw: Pax, 1962, volume 1.

John Paul II, Pope. *Uniwersytety w nauczaniu Jana Pawła II* [*Universities in the Teaching of John Paul II*]. No editor given. Warsaw: Szkoła Wyższa im. Bogdana Jańskiego [Bogdan Janski Higher School], 1999, volume 1.

Jaroszyński, Piotr. *Metafizyka i sztuka* [*Metaphysics and Art*]. Warsaw: Gutenberg-Press, 1996.

————. "Spór o przedmiot 'Metafizyki' Arystotelesa" ["Dispute about the Object of Aristotle's 'Metaphysics'"], in *Roczniki Filozoficzne* [*Philosophical Yearbooks*]. Edited by Towarzystwo Nauknowe, 31:1 (1983).

Kamiński, Stanisław. *Pojęcie nauki i klasyfikacja nauk* [*Concept of Science and the Classification of the Sciences, The*]. Lublin: Catholic University of Lublin, 1981.

Kingsley, Peter. *Ancient Philosophy, Mystery, and Magic: Empedocles and Pythagorean Tradition.* Oxford: The Clarendon Press, 1995.

Krąpiec, Mieczysław Albert. *Byt i istota [Being and Essence].* Lublin: Redacja Wydawnictw [Editorial Board of Publishers], Catholic University of Lublin, 1994.

―――. *Filozofia—co wyjaśnia? [Philosophy—What does it Explain?].* Lublin: Lubelska Szkoła Filozofii Chrześcijańskiej [Lublin School of Christian Philosophy], 1999.

―――. *Filozofia w teologii [Philosophy in Theology],* Lublin, Instytut Edukacji Narodowej, 1998.

―――. *Ja-człowiek [I-man].* Lublin: Redacja Wydawnictw [Editorial Board of Publishers], Catholic University of Lublin, 1991.

―――. *O ludzką politykę [On a Human Politics].* Warsaw: Guttenberg-Press, 1996.

―――. *Realizm ludzkiego poznania [Realism of Human Cognition].* Poznań: Pallotinum, 1959.

―――, Stanisław Kamiński. *Z teorii i metodologii metafizyki [On the Theory and Methodology of Metaphysics].* Lublin, Redacja Wydawnictw [Editorial Board of Publishers], Catholic University of Lublin, 1994.

Martin, Hans-Peter and Harald Schumann. *Pułapka globalizacji. Atak na demokrację i dobrobyt [The GlobalizationTrap: Attack on Democracy and Prosperity].* Translated into English by Patrick Camiller, New York: Zed Books, 1997; translated into Polish by M. Zybura. Original German title: *Globaliserungsfalle.* Wrocław: Wydawnictwo Dolnośląskie, 1999.

Nauknowe, Towarzystwo (ed.). *Roczniki Filozoficzne [Philosophical Yearbooks],* 31:1 (1983).

Pigulewska, Nina W.. *Kultura syryjska we wczesnym średniowieczu [Syrian Culture in the Early Middle Ages].* Warsaw: Pax, 1989.

Quispel, Giles. *Gnoza [Gnosticism].* Translated by B. Kita. Warsaw: Pax, 1988.

Rossi, Pado. *Filozofowie i maszyny [Philosophers and Machines].* Translated by A. Kreisberg. Warsaw: Państwowe Wydawnictwo Naukowe [State Scientific Publisher], 1978.

"Rozum otwarty na wiarę. 'Fides et ratio'—w rocznicę ogłoszenia. II Międzynarodowe Sympozjum Metafizyczne, KUL, 9-10.XII.1999" ["Reason Open to Faith. *'Fides et ratio'*—on the Anniversary of the Proclamation. Second International Metaphysical Symposium, Catholic University of Lublin, December 9–10, 1999"]. Lublin: *Polskie Towarzystwo Tomasza z Akwinu* [Polish Society of Thomas Aquinas], 2000.

Scheler, Max. *Pisma z antropologii filozoficznej i teorii wiedzy* [*Writings in Philosophical Anthropology and the Theory of Knowledge*]. Translated by StanisławCzerniak and, A. Węgrzycki, Warsaw: Państwowe Wydawnictwo Naukowe [State Scientific Publisher], 1987.

Sprague de Camp, Lyon. *Wielcy i mali twórcy cywilizacji* [*Major and Minor Creators of Civilization*]. Translated by B. Orłowski, Warsaw: Wiedza Powszechna, 1968.

Tatarkiewicz, Władysław. *Historia filozofii* [*History of Philosophy*]. Warsaw: Państwowe Wydawnictwo Naukowe [State Scientific Publisher]. 1970, volumes 1–3.

Weber, Max. *Etyka protestancka a duch kapitalizmu* [*The Protestant Ethic and the Spirit of Capitalism*]. Translated by Jan Miziński, Lublin: Test, 1994.

Wielgus, Stanisław . "Otwarcie Społecznego Tygodnia na jubileusz Metropolii Warszawskiej" ["Opening of the Social Week for the Jubilee of the Warsaw Metropolis"] (May 18, 2000), in *Człowiek w kulturze* [*Man and Culture*]. Lublin: Catholic University of Lublin, 2002, volume 14.

Woroniecki, Jacek W.. *Katolicka etyka wychowawcza* [*Catholic Educational Ethics*]. Lublin: Redacja Wydawnictw [Editorial Board of Publishers], Catholic University of Lublin, 1986, volume 1.

ABOUT THE AUTHOR

Piotr Jaroszyński (born 1955) holds the Chair of the Philosophy of Culture at the John Paul II Catholic University of Lublin, Poland. He works within the framework of classical philosophy. He belongs to the American Catholic Philosophical Association, American Maritain Association, Gilson Society and the International Society of Thomas Aquinas. He has written many articles and several scholarly books, including *The Controversy over Beauty* (1992, in Polish), *Metaphysics and Art* (2002), *Ethics: the Drama of the Moral Life* (2003). For the last book he received the personal thanks of Pope John Paul II.

INDEX OF WORKS

Peter A. Redpath prepared all indices

Page numbers refer to citation of works in the body of text, notes, and bibliography

Note: Numbers indicated in *boldface italics* refer to pages with Illustrations

Index of Works 283

INDEX OF AUTHORS, EDITORS, AND TRANSLATORS

This index lists persons in their capacity as authors, editors, and translators
and as individuals named anywhere in this monograph

Note: Numbers indicated in *boldface italics* refer to pages with Illustrations

Keyes, Clinton Walker, 233, 246, 260

Kingsley, Peter, 247, 248, 272

Kita, B., 251, 272

Klemm, Friedrich, 245, 267

Kornatowski, Wictor, 263

Krąpiec, Mieczysław A., ix, xiii, xv, *1*, 38, *75*, 98, 161, 162, 194, 195, 229, 233, 234, 236, 237, 239, 241, 243, 244, 245, 250, 251, 253, 254, 256, 257, 262, 272

Kreisberg, A. 272

Krokiewicz, Adam, 240, 264

Kwiatkowski, Tadeusz, ix, xvii

Lafleur, Laurence J., 261, 279

Lamb, Walter R. M., 244, 261

Landry B., 256, 260

Largeault, Jean, 245, 268

Larrain, Jorge, 253, 267

Layton B., 251, 265, 267, 269

Le Doeuff, Michèle, 249, 262

Leibniz, Gottfried Wilhelm von, 146, 147, 154, 155, 159, 168, 185, 248, 264

Leśniak, Kazimierz, 111, 234, 244, 245, 262

Lewy, Hans, 72, 240, 268

Lindberg, David C., 243, 244, 245, 248, 249, 251, 265, 266, 268

Llasera, Margeret, 249, 260

Locke, John, 154, 179, 181, 183, 184, 185, 187, 252, 253, 261

Luther, Martin, 134, 147, 162, 170, 247, 268

Łypacewicz , S., 243, 244, 271

Machiavelli, Niccoló, 162

Malherbe, Michael, 250, 268

Mandonnet, Pierre, 242, 259

Manuel, Frank E., 249, 268

Manuel, Fritzie P., 249, 268

Marietti, 242, 243, 259

Maritain, Jacques, 247, 249, 268, 269, 275

Martin, Hans-Peter, 256, 273

Maryniarczyk, Andrzej, 233, 234, 236, 237, 238, 239, 241, 242, 244, 257, 263

Mates, Benson, 238, 268

Mathias, Peter, 246, 247, 249, 250, 270

Maurer, Armand A., 252, 266, 268

Maykowska, Maria, 237, 263

McDonald, Hugh, v, xiii, xxi, *51*, *175*, *191*, *217*, 256, 268

McKeon, Richard, 236, 268

McMullin Ernan, 6, 233, 249, 268

Meiklejohn, J. M. D., 253, 256, 261

Mettler, Dieter, 249, 264, 267

Migne, 242, 262

Merlan, Philippe, 235, 245, 268

Milton, John, 130, 246

Miziński, Jan, 255, 273

Mojsisch, Zob B., 251, 268

Montloin, Pierre, 151, 249, 268

Moos, M. F., 242, 259

More, Thomas, St., 149, 263

Moss, L. B., 243, 264

Mroczkowska, Z., 238, 272

Myszor, Wincenty, 251

Nardi, Bruno, 241, 260

Nauknowe, Towarzystwo, 235, 272, 273

Newton, Edward, 256, 268

Newton, Sir Isaac, 153, 154, 159, 180, 196

Orłowski, B., 245, 273

Orwell, George, 214, 215

Owens, Joseph, 36, 157, 235, 236, 243, 245, 250, 269

Peltonen, Markku, 246, 248, 250, 269, 270

Peter Lombard, 88, 89, 90

Peuckert, Will-Eric, 251

Pianko, Gabriela, 262

Pico della Mirandola, Giovanni, 143, 144, 146, 248

Pigulewska, Nina W., 243, 273

Piotrowicz, Lodovici, 234, 246, 262

INDEX OF SUBJECTS

INDEX OF NAMES

Note: Numbers indicated in **_boldface italics_** refer to pages with Illustrations

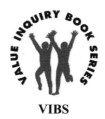

VIBS

The **Value Inquiry Book Series** is co-sponsored by:

Adler School of Professional Psychology
American Indian Philosophy Association
American Maritain Association
American Society for Value Inquiry
Association for Process Philosophy of Education
Canadian Society for Philosophical Practice
Center for Bioethics, University of Turku
Center for Professional and Applied Ethics, University of North Carolina at Charlotte
Central European Pragmatist Forum
Centre for Applied Ethics, Hong Kong Baptist University
Centre for Cultural Research, Aarhus University
Centre for Professional Ethics, University of Central Lancashire
Centre for the Study of Philosophy and Religion, University College of Cape Breton
Centro de Estudos em Filosofia Americana, Brazil
College of Education and Allied Professions, Bowling Green State University
College of Liberal Arts, Rochester Institute of Technology
Concerned Philosophers for Peace
Conference of Philosophical Societies
Department of Moral and Social Philosophy, University of Helsinki
Gannon University
Gilson Society
Haitian Studies Association
Ikeda University
Institute of Philosophy of the High Council of Scientific Research, Spain
International Academy of Philosophy of the Principality of Liechtenstein
International Association of Bioethics
International Center for the Arts, Humanities, and Value Inquiry
International Society for Universal Dialogue
Natural Law Society
Philosophical Society of Finland
Philosophy Born of Struggle Association
Philosophy Seminar, University of Mainz
Pragmatism Archive at The Oklahoma State University
R.S. Hartman Institute for Formal and Applied Axiology
Research Institute, Lakeridge Health Corporation
Russian Philosophical Society
Society for Existential Analysis
Society for Iberian and Latin-American Thought
Society for the Philosophic Study of Genocide and the Holocaust
Unit for Research in Cognitive Neuroscience, Autonomous University of Barcelona
Yves R. Simon Institute

Titles Published

1. Noel Balzer, *The Human Being as a Logical Thinker*

2. Archie J. Bahm, *Axiology: The Science of Values*

3. H. P. P. (Hennie) Lötter, *Justice for an Unjust Society*

4. H. G. Callaway, *Context for Meaning and Analysis: A Critical Study in the Philosophy of Language*

5. Benjamin S. Llamzon, *A Humane Case for Moral Intuition*

6. James R. Watson, *Between Auschwitz and Tradition: Postmodern Reflections on the Task of Thinking.* A volume in **Holocaust and Genocide Studies**

7. Robert S. Hartman, *Freedom to Live: The Robert Hartman Story,* Edited by Arthur R. Ellis. A volume in **Hartman Institute Axiology Studies**

8. Archie J. Bahm, *Ethics: The Science of Oughtness*

9. George David Miller, *An Idiosyncratic Ethics; Or, the Lauramachean Ethics*

10. Joseph P. DeMarco, *A Coherence Theory in Ethics*

11. Frank G. Forrest, *Valuemetrics^x: The Science of Personal and Professional Ethics.* A volume in **Hartman Institute Axiology Studies**

12. William Gerber, *The Meaning of Life: Insights of the World's Great Thinkers*

13. Richard T. Hull, Editor, *A Quarter Century of Value Inquiry: Presidential Addresses of the American Society for Value Inquiry.* A volume in **Histories and Addresses of Philosophical Societies**

14. William Gerber, *Nuggets of Wisdom from Great Jewish Thinkers: From Biblical Times to the Present*

60.   Palmer Talbutt, Jr., Rough Dialectics: *Sorokin's Philosophy of Value*, with contributions by Lawrence T. Nichols and Pitirim A. Sorokin

61.   C. L. Sheng, *A Utilitarian General Theory of Value*

62.   George David Miller, *Negotiating Toward Truth: The Extinction of Teachers and Students.* Epilogue by Mark Roelof Eleveld. A volume in **Philosophy of Education**

63.   William Gerber, *Love, Poetry, and Immortality: Luminous Insights of the World's Great Thinkers*

64.   Dane R. Gordon, Editor, *Philosophy in Post-Communist Europe.* A volume in **Post-Communist European Thought**

65.   Dane R. Gordon and Józef Niznik, Editors, *Criticism and Defense of Rationality in Contemporary Philosophy.* A volume in **Post-Communist European Thought**

66.   John R. Shook, *Pragmatism: An Annotated Bibliography, 1898-1940.* With contributions by E. Paul Colella, Lesley Friedman, Frank X. Ryan, and Ignas K. Skrupskelis

67.   Lansana Keita, *The Human Project and the Temptations of Science*

68.   Michael M. Kazanjian, *Phenomenology and Education: Cosmology, Co-Being, and Core Curriculum.* A volume in **Philosophy of Education**

69.   James W. Vice, *The Reopening of the American Mind: On Skepticism and Constitutionalism*

70.   Sarah Bishop Merrill, *Defining Personhood: Toward the Ethics of Quality in Clinical Care*

71.   Dane R. Gordon, *Philosophy and Vision*

72.   Alan Milchman and Alan Rosenberg, Editors, *Postmodernism and the Holocaust.* A volume in **Holocaust and Genocide Studies**

73.   Peter A. Redpath, *Masquerade of the Dream Walkers: Prophetic Theology from the Cartesians to Hegel.* A volume in **Studies in the History of Western Philosophy**

89.   Yuval Lurie, *Cultural Beings: Reading the Philosophers of Genesis*

90.   Sandra A. Wawrytko, Editor, *The Problem of Evil: An Intercultural Exploration*. A volume in **Philosophy and Psychology**

91.   Gary J. Acquaviva, *Values, Violence, and Our Future*. A volume in **Hartman Institute Axiology Studies**

92.   Michael R. Rhodes, *Coercion: A Nonevaluative Approach*

93.   Jacques Kriel, *Matter, Mind, and Medicine: Transforming the Clinical Method*

94.   Haim Gordon, *Dwelling Poetically: Educational Challenges in Heidegger's Thinking on Poetry*. A volume in **Philosophy of Education**

95.   Ludwig Grünberg, *The Mystery of Values: Studies in Axiology*, Edited by Cornelia Grünberg and Laura Grünberg

96.   Gerhold K. Becker, Editor, *The Moral Status of Persons: Perspectives on Bioethics*. A volume in **Studies in Applied Ethics**

97.   Roxanne Claire Farrar, *Sartrean Dialectics: A Method for Critical Discourse on Aesthetic Experience*

98.   Ugo Spirito, *Memoirs of the Twentieth Century*. Translated from Italian and Edited by Anthony G. Costantini. A volume in **Values in Italian Philosophy**

99.   Steven Schroeder, *Between Freedom and Necessity: An Essay on the Place of Value*

100.   Foster N. Walker, *Enjoyment and the Activity of Mind: Dialogues on Whitehead and Education*. A volume in **Philosophy of Education**

101.   Avi Sagi, Kierkegaard, *Religion, and Existence: The Voyage of the Self*. Translated from Hebrew by Batya Stein

102.   Bennie R. Crockett, Jr., Editor, *Addresses of the Mississippi Philosophical Association*. A volume in **Histories and Addresses of Philosophical Societies**

130. Richard Rumana, *Richard Rorty: An Annotated Bibliography of Secondary Literature*. A volume in **Studies in Pragmatism and Values**

131. Stephen Schneck, Editor, *Max Scheler's Acting Persons: New Perspectives* A volume in **Personalist Studies**

132. Michael Kazanjian, *Learning Values Lifelong: From Inert Ideas to Wholes*. A volume in **Philosophy of Education**

133. Rudolph Alexander Kofi Cain, Alain Leroy Locke: *Race, Culture, and the Education of African American Adults*. A volume in **African American Philosophy**

134. Werner Krieglstein, *Compassion: A New Philosophy of the Other*

135. Robert N. Fisher, Daniel T. Primozic, Peter A. Day, and Joel A. Thompson, Editors, *Suffering, Death, and Identity*. A volume in **Personalist Studies**

136. Steven Schroeder, *Touching Philosophy, Sounding Religion, Placing Education*. A volume in **Philosophy of Education**

137. Guy DeBrock, *Process Pragmatism: Essays on a Quiet Philosophical Revolution*. A volume in **Studies in Pragmatism and Values**

138. Lennart Nordenfelt and Per-Erik Liss, Editors, *Dimensions of Health and Health Promotion*

139. Amihud Gilead, *Singularity and Other Possibilities: Panenmentalist Novelties*

140. Samantha Mei-che Pang, *Nursing Ethics in Modern China: Conflicting Values and Competing Role Requirements*. A volume in **Studies in Applied Ethics**

141. Christine M. Koggel, Allannah Furlong, and Charles Levin, Editors, *Confidential Relationships: Psychoanalytic, Ethical, and Legal Contexts*. A volume in **Philosophy and Psychology**

142. Peter A. Redpath, Editor, *A Thomistic Tapestry: Essays in Memory of Étienne Gilson*. A volume in **Gilson Studies**

157.   Javier Muguerza, *Ethics and Perplexity: Toward a Critique of Dialogical Reason*. Translated from the Spanish by Jody L. Doran. Edited by John R. Welch. A volume in **Philosophy in Spain**

158.   Gregory F. Mellema, *The Expectations of Morality*

159.   Robert Ginsberg, *The Aesthetics of Ruins*

160.   Stan van Hooft, *Life, Death, and Subjectivity: Moral Sources in Bioethics* A volume in **Values in Bioethics**

161.   André Mineau, *Operation Barbarossa: Ideology and Ethics Against Human Dignity*

162.   Arthur Efron, *Expriencing Tess of the D'Urbervilles: A Deweyan Account.* A volume in **Studies in Pragmatism and Values**

163.   Reyes Mate, *Memory of the West: The Contemporaneity of Forgotten Jewish Thinkers*. Translated from the Spanish by Anne Day Dewey. Edited by John R. Welch. A volume in **Philosophy in Spain**

164.   Nancy Nyquist Potter, Editor, *Putting Peace into Practice: Evaluating Policy on Local and Global Levels*. A volume in **Philosophy of Peace**

165.   Matti Häyry, Tuija Takala, and Peter Herissone-Kelly, Editors, *Bioethics and Social Reality*. A volume in **Values in Bioethics**

166.   Maureen Sie, *Justifying Blame: Why Free Will Matters and Why it Does Not*. A volume in **Studies in Applied Ethics**

167.   Leszek Koczanowicz and Beth J. Singer, Editors, *Democracy and the Post-Totalitarian Experience*. A volume in **Studies in Pragmatism and Values**

168.   Michael W. Riley, *Plato's* Cratylus: *Argument, Form, and Structure*. A volume in **Studies in the History of Western Philosophy**

169.   Leon Pomeroy, *The New Science of Axiological Psychology*. Edited by Rem B. Edwards. A volume in **Hartman Institute Axiology Studies**

170.   Eric Wolf Fried, *Inwardness and Morality*

171.   Sami Pihlstrom, *Pragmatic Moral Realism: A Transcendental Defense.* A volume in Studies in **Pragmatism and Values**